Population Dynamics and Projection Methods

Understanding Population Trends and Processes

Volume 4

Series Editor

J. Stillwell

In western Europe and other developed parts of the world, there are some very significant demographic processes taking place at the individual, household, community and national scales including the ageing of the population, the delay in childbearing, the rise in childlessness, the increase in divorce, the fall in marriage rates, the increase in cohabitation, the increase in mixed marriages, the change in household structures, the rise in step-parenting and the appearance of new streams of migration taking place both within and between countries. The relationships between demographic change, international migration, labour and housing market dynamics, care provision and intergenerational attitudes are complex to understand and yet it is vital to quantify the trends and to understand the processes. Similarly, it is critical to appreciate what the policy consequences are for the trends and processes that have become apparent. This series has its roots in understanding and analysing these trends and processes.
This series will be of interest to a wide range of individuals concerned with demographic and social change, including demographers, population geographers, sociologists, economists, political scientists, epidemiologists and health researchers as well as practitioners and commentators across the social sciences.

For further volumes:
http://www.springer.com/series/8113

Population Dynamics and Projection Methods

Understanding Population Trends and Processes – Volume 4

Edited by

John Stillwell
University of Leeds, UK

and

Martin Clarke
University of Leeds, UK

Springer

Editors
Prof. John Stillwell
University of Leeds
School of Geography
Woodhouse Lane
LS2 9JT Leeds
UK
j.c.h.stillwell@leeds.ac.uk

Prof. Martin Clarke
University of Leeds
School of Geography
Woodhouse Lane
LS2 9JT Leeds
UK
m.c.clarke@leeds.ac.uk

ISBN 978-90-481-8929-8 e-ISBN 978-90-481-8930-4
DOI 10.1007/978-90-481-8930-4
Springer Dordrecht Heidelberg London New York

Library of Congress Control Number: 2011920950

Printed on acid-free paper

Springer is part of Springer Science+Business Media (www.springer.com)

Preface

Good understanding of contemporary demographic structure and population dynamics underpins effective planning and decision making for the future. One of the key contributors to the development of a range of population projection methodologies over the past 40 years is Professor Philip Rees at the University of Leeds. This book contains an eclectic range of methodological and substantive contributions by a number of eminent researchers in the fields of population geography and demography that were presented at a symposium in July 2009 to honour Philip Rees' retirement. Different macro and micro approaches for estimating and projecting populations are reviewed, trends in the components of change in the UK (births, deaths and migration) are presented, international comparisons of internal migration are drawn, impacts of population ageing are considered and a new perspective on understanding urban evolution is offered. All these themes are interconnected in one way or another but the key dimension of linkage in this particular volume is that, collectively, they represent a compendium of Phil's research interests and the celebration of a lifetime of commitment to undertaking meticulous analytical research, developing innovative modelling methods and enhancing knowledge in population geography and spatial demography.

Demographers have always been interested in change and have wrestled with the demands of policy makers to inform them about the future of national and sub-national populations and their productive capability or destructive potential. Chapter 1 of the book, by *Philip Rees* himself (Fig. 1), considers the dynamics of global population change by reviewing the projections of world population development over the next 90 years and the regional differences that will become apparent under the assumptions adopted by the major projection agencies. The key factors influencing the ageing of most national populations include low or declining fertility (until relatively recently), rising life expectancy and international migration gains and losses. These demographic processes are creating older and more diverse populations. Within national borders, internal migration is the main driver for redistributing these changing populations. Drawing on recent research on ethnic group projections (Rees, Wohland, Norman, & Boden, 2010), undertaken as part of the 'Understanding Population Trends and Processes (UPTAP)' initiative and funded by the Economic and Social Research Council (ESRC) in the UK, Rees compares

Fig. 1 Phil Rees delivering his lecture at the symposium on 1 July 2009

alternative projections of ethnic fertility in the UK and then presents new estimates of life expectancy by ethnic group in the UK which he contrasts with experience in the USA. The final sections of the chapter provide us with some indications of how the ethnic composition of the UK will change between 2001 and 2051 as minority groups mature and as ageing exerts its influence on the demographic structure of all groups within the population. Whilst the demographic changes identified in this chapter have very significant policy implications, not least for the provision of services, many policy questions need more than demographic explanations; there is a requirement also to understand changes in educational attainment, participation in the labour force, family and household formation (many demands are household generated), changes in retirement dependent on state support and savings and on health/illness outcomes, all of which are underpinned by the availability of population projections.

A variety of strategies are available to model our changing populations. Sometimes population stocks are modelled as a time series (such as when a consultancy project needs swift execution). However, it is better to study and model transitions between states. The classic cohort-component model is still at the heart of the projection of large and small populations, in many forms as single region, bi-regional or multi-region models, whether implemented using macro-populations experiencing average group intensities or using micro-populations of individuals sampling their transitions from a macro-distribution. What is vital in implementing such models is that the input intensities match model intensities in definition. Even the best demographers can get this wrong. Sometimes the input data are inadequate and it becomes necessary to resort to other modelling strategies. To solve the student migration issue (poor data on the migration of students on graduation), recent work has introduced agent based models into microsimulation, for example (Wu, Birkin, & Rees, 2008). In other situations, no direct estimates of inputs to projection models are available and the needed inputs must be modelled from proxy variables, survey data or administrative sources, as has been done recently for ethnic group mortality, fertility and immigration in the UK (Norman, Rees, Boden, & Wohland, 2010).

To drive our projection models, assumptions must be made and uncertainty must be dealt with. Traditionally high and low variants are used, but will be replaced in the future by stochastic projections. These use time series models or historical analysis or expert opinion to produce error distributions of the main projection drivers for sampling. It is also necessary to think 'out of the box' and develop scenarios that establish 'what might happen if' such as: What if the conventional view of continuous life expectancy improvement was replaced by predictions of higher mortality from the obesity epidemic? What if climate change crises caused a rise in the migration of environmental migrants? What if resource depletion (the end of oil) reduced mobility and migration in Europe? These are scenarios which the European Commission has charged a network of European demographers to explore through a major ESPON project (de Beer et al., 2010).

In the 1970s, Philip Rees pioneered multistate demographic accounts involving tables of flows that are moves/transitions between states. The population covered by an account, the age-time framework, the observation window and the states distinguished define the boundary of an account which provides a framework for the measurement of flows and the estimation of flows when data are incomplete. An account-based model provides a framework for combining flows and stocks in a consistent manner. A major aim in the development of accounts is that the data are valid (i.e. measure what they are supposed to measure), reliable and timely. Chapter 2 by *Frans Willekens* presents the major principles of multistate demographic accounting developed by Philip Rees and extended by his team and others. It incorporates some aspects of multistate modelling in survival analysis. The account that results includes population flows, population stocks and durations of exposure and is a basis for the estimation of transition rates and probabilities for demographic modelling (life tables and projections).

It is well-known that existing models of population projection, including those that have multiregional structures, do not adequately handle the international migration component of population dynamics. In many cases, they are not adapted to the modelling of the populations of several countries as well as regions within those countries simultaneously. The arguments for taking international migration into account in the modelling of population change are twofold. First, population processes are systemic in nature and international migration is an interaction between elements of the population system, that is between national and regional populations of various countries, that should be included since the volume of international migration has been growing in the last 20 years. Second, there is a very practical argument that the forecasting errors arising directly from ignoring international migration are very large; Rees, Kupiszewski, Eyre, Wilson, and Durham (2001) have shown that the magnitude of errors using Eurostat forecasts for the 1980s. Thus, the incorporation of international migration into models of population dynamics is a key issue in terms of reducing forecasting error and one way to achieve this is to use a matrix of flows between countries, rather than net migration, and to use emigration rates where possible. These developments have led *Marek Kupiszewski* and *Dorota Kupiszewska* to use the *ECPOP* model (Rees, Stillwell, & Convey, 1992; Rees, 1996) as the basis for constructing *MULTIPOLES*, a multiregional multilevel model of population dynamics which takes into account international migration. *MULTIPOLES*, which simultaneously models population change in countries and regions and takes into account international migration between as well as from the 'rest of the world', is explained in Chapter 3 where an application of the model to forecast the elderly population of countries in central and eastern Europe is also presented.

Chapter 4 by *Tom Wilson* describes the model, assumptions and projection outputs from the official New South Wales Government 2008 release population projections. The model, the *New* South Wales *D*emographic *S*imulation *S*ystem (*NEWDSS*), incorporates directional migration modelling and produces projections at three geographical scales: (i) New South Wales and the rest of Australia; (ii) major regions of the State; and (iii) Statistical Local Areas. The system utilises movement accounts-based projection models at the State and regional scales and a transition accounts-based model at the local area scale. One of the innovative feature of *NEWDSS* is the way the local area transition accounts-based model uses migration probabilities based on census data but, to simplify assumption-setting, constrains the projections to net movement assumptions. Projected population accounts at this scale are then presented in the form of movement accounts to ease understanding by non-technical users. The chapter describes the principal aspects of the model, provides an overview of how the projection assumptions were prepared and discusses some of practical issues which arise in preparing local area projections. Key aspects of the demographic future of New South Wales for the period 2006–2036 are presented.

In contrast to the first five chapters of the book which outline alternative approaches to projection and demonstrate projection methods used in different contexts, the next three chapters concentrate on specific components of population change: fertility, mortality and migration. In Chapter 6, *Paul Norman* focuses on the relationship between fertility and infant mortality in the United Kingdom and the hypothesis that reductions in fertility are a direct result of falls in infant mortality. Whilst William Brass found little evidence at regional and county level in England and Wales of changing geographies of fertility and child mortality between 1876 and 1928, with no detectable 'direct influence of child mortality on fertility' (Brass & Kabir, 1979, p. 86), recent studies of the late twentieth and early twenty-first centuries show distinct geographic variations in both fertility (Boyle, 2003; Boyle et al., 2007; Tromans et al., 2008) and infant mortality (Norman et al., 2008). Brass' study framework has been adopted in Chapter 6 for analysis using UK-wide data for local authorities for the period 1981–2006. The results suggest that the relationship between trends in infant mortality and fertility remain unclear and that each of these indicators is influenced by different variables.

Migration is the most uncertain component of population change and the one which requires careful monitoring. The problem is that UK migration statistics, be they on internal or international migration flows, are unacceptably poor for use in monitoring and policy making. There has been a significant outcry for improvement and the Government has responded by establishing programmes of actions and activities to meet the requirements for better data. In this context, the aim of Chapter 6 by *John Stillwell, Peter Boden* and *Adam Dennett*, is to review the need for migration statistics and the current predicament over reliable data and then to illustrate some examples of migration information systems that have been developed in an academic environment for different type of users so as to support the monitoring of migration trends over time and better analysis of the changing patterns and complexion of migration.

Cross-national comparisons of demographic indices provide valuable insights into the status and trajectory of different societies. However, whilst demographic indicators such as total fertility rate, life expectancy or total immigration rate are relatively easy to calculate from available data for many countries of the world, indicators of internal migration intensity that are comparable between nations prove more difficult to compute. This is the case for a number of reasons, as *Martin Bell* and *Salut Muhidin* establish in Chapter 7, including differences in the way internal migration is defined, in the time intervals over which internal migration is measured and, in particular, in the way in which national territories are divided spatially in different countries. Every country has its own hierarchy of geographical areas and because these areas differ in areal or population size, it is difficult to make accurate comparisons between countries of a phenomenon that involves interaction between and within zones in different tiers of the spatial hierarchy. It is very unlikely that countries have zones that are similar in size and shape, between or within which the volume of migration can be compared directly. It is for these reasons that Bell and Muhidin turn to an indicator introduced by the French demographer, Daniel Courgeau, in order to compare internal migration across 27 countries from Asia, Africa, Latin America and the Caribbean as well as the developed world. Courgeau's *k* is a synthetic measure which indicates internal migration intensities at a range of spatial scales since it is the slope of the regression line that connects the migration intensities involved. The results are intriguing and provide evidence that migration intensities in countries of the developed world tend to be relatively high whereas Asian countries tend to be at the other end of the mobility spectrum. When Courgeau's *k* values for different countries are compared over time, increasing intensities are apparent when the lifetime migration data are used and declining trends are evident from the 5-year migration intensities. Thus, the analysis tells us that whilst more people are moving from their place of birth during their lifetime, the trend in intensity of movement in the last three decades based on 5-year data is downwards.

In Chapter 8, *Les Mayhew* addresses some of the major issues around demographic restructuring that are currently confronting policy makers as they try to establish what the implications of an ageing population will be over the next two decades. Whilst increased longevity is a laudable aim, we must be conscious that this implies greater and greater need for health care to maintain healthy life expectancy whilst simultaneously a sharp decline in the old age support ratio implies longer and longer working lives for the healthy population of working age. This chapter is concerned with explaining these trends and the relationship between ageing, health and work so that the implications of getting the outcomes wrong are better understood as well as the potential economic benefits of getting better outcomes in later life right through focusing on better health and greater participation. The chapter uses three concepts in particular: life expectancy; healthy life expectancy; and working life expectancy. It explains what these concepts are and considers optimistic or pessimistic hypotheses that follow from changes in the quantities of these concepts. An 'active ageing' scenario, for example, which narrows the gap between life expectancy and healthy life expectancy and which increases working life expectancy, is to be welcomed as it will improve living standards and reduce the need for immigration.

The focus of most chapters of the book is on macro-demographic theories, methods and applications with consideration for particular components of population change or specific sets of sub-national populations. However, there is a huge body of work that has evolved over the last 60 or so years that is based on a building micro models of individual (e.g. person or household) behaviour. In Chapter 9, *Mark Birkin* and *Martin Clarke* provide a review of these approaches by tracing the development of microsimulation models from their origins in the late 1950s through to the present time, noting how it was only in the 1970s that geographers (including Phil Rees) embraced the approach and began exploring the application of the models in a spatial setting. They examine issues around population reconstruction within spatial microsimulation, outline some of the issues surrounding household dynamics and consider some of the new developments taking place involving the development of agent-based models and attempts to embed behaviour into these models.

In the final chapter of the book, the focus moves to the evolution of cities as nonlinear dynamical systems. It is known that, in general, for such systems, the step from one period to the next is highly dependent on the initial conditions prevailing at the beginning of the period. The evolution of a city is a sequence of such steps and this kind of evolution is said to be path dependent. It is shown that at any one time, the initial conditions for a city can be characterised by an analogue of DNA. The argument can be applied to a system of cities as well as to a particular city. Urban history, therefore, can be seen as giving an account of the evolution of this DNA in either the intra-urban or the inter-urban case. In this chapter, *Alan Wilson* and *Joel Dearden* re-interpret the urban retail model as a model of a system of cities but also with an emphasis on the impact of the evolution of transport systems. An appropriate model is articulated and illustrated by the growth of Chicago in the nineteenth century which offers a new perspective for historical geography and can also be seen as a novel way of modelling population dynamics.

The symposium in July 2009 to celebrate Phil's retirement was attended by past and present colleagues from far and wide (Fig. 2). It is fitting to draw this short introduction to an end with a short poem entitled 'Working with Phil' that was penned at the time by Nicole van der Gaag from the Netherlands Interdisciplinary Institute, an organisation with which Phil has had close association over many years. This poem expresses many of the sentiments of Phil's friends and collaborators.

Working with Phil

Geography, countries, but most of all regions
All over Europe, from NUTS 1 to NUTS 5
The continuous search for regional data
You could easily drown in, but with Phil you'll survive

To migration and models, with figures and arrows
To examples of flows from part A to part B
To mobility, maps, even poems of Shakespeare
To workshops and papers and dinners and tea

It also reminds me of sea-level stories
To the border between Holland and the UK
To what happens if sea level ceaselessly rises
To the parts of our countries that will fade away

I know Phil as an expert in several topics
The projects he touches they turn into gold
He's always so modest and never pretending
For persons he works with a keystone to hold

Top of all that he is, he's a very nice person
Kind-hearted and gentle, and always in peace
I'm honoured and grateful for our collaboration
It was ever a pleasure, I felt always at ease.

Nicole van der Gaag, July 2010

Leeds, UK

John Stillwell
Martin Clarke

Fig. 2 Participants attending the symposium on 1 July 2009

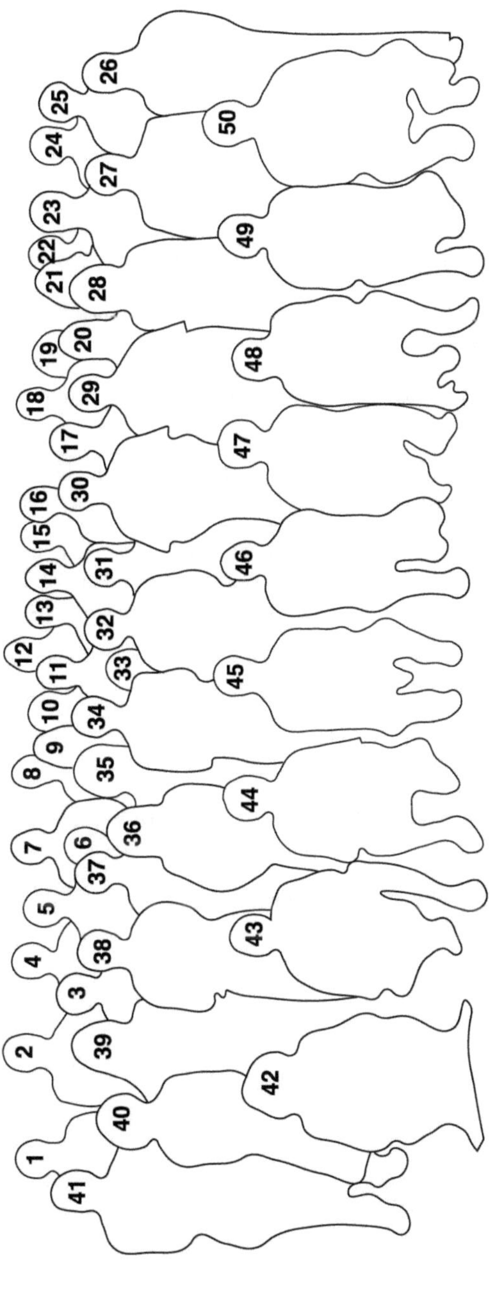

1 Myles Gould
2 Mark Birkin
3 Frans Willekens
4 Robin Butlin
5 Les Mayhew
6 Angela Dale
7 Bob Woods
8 Martin Bell
9 Pia Wohland
10 Nicole van der Gaag
11 Cecilia MacIntyre
12 Leo van Wissen
13 Paul Norman
14 Nico Keilman
15 Graham Clarke
16 Maryvonne Plessis-Fraissard
17 Martyn Senior
18 John McCarthy
19 Helen Durham
20 Belinda Wu
21 Linda See
22 Heather Eyre
23 John Jenkins
24 Marek Kupiszewski
25 Dorota Kupiszewska
26 Caroline Hoy
27 James Raymer
28 Lisa Youngman
29 Dan Vickers
30 John Stillwell
31 Darren Smith
32 Phil Rees
33 Peter Goldblatt
34 Sir Alan Wilson
35 Julia Williams
36 Rukchanok Karcharnubarn
37 Adrian McDonald
38 Peter Boden
39 Maja Biernacka
40 Beata Nowok
41 Alex Hirschfield
42 Dimitris Ballas
43 Patrick Sim
44 Paul Williamson
45 Adam Dennett
46 Dieter Kramer
47 Jianhui Jin
48 Oliver Duke-Williams
49 Andy Peloe
50 Tom Wilson

References

Boyle, P. J. (2003). Population geography: Does geography matter in fertility research? *Progress in Human Geography*, *27*(5), 615–626.

Boyle, P. J, Graham, E., & Feng, Z. (2007). *Contextualising demography: The significance of local clusters of fertility in Scotland*. Working Paper 2007-036, Max Planck Institute for Demographic Research, Rostock.

Brass, W., & Kabir, M. (1979) Regional variations in fertility and child mortality during the demographic transition in England and Wales. In J. Hobcraft & P. Rees (Eds.), *Regional demographic development* (pp. 71–88). London: Croom Helm.

de Beer, J., van der Gaag, N., van der Erf, R., Bauer, R., Fassmann, H., Kupiszewska, D., et al. (2010). *DEMIFER, Demographic and migratory flows affecting European regions and cities*, Applied Research Project 2013/1/3, Final Report, The ESPON 2013 Programme. Luxembourg: ESPON and The Netherlands: NIDI.

Norman, P., Gregory, I., Dorling, D., & Baker, A. (2008). Geographical trends in infant mortality: England and Wales, 1970–2006. *Health Statistics Quarterly*, *40*, 18–29.

Norman, P., Rees, P., Boden, P., & Wohland, P. (2010). Ethnic group populations: The components for projection, demographic rates and trends. In J. Stillwell & M. van Ham (Eds.), *Ethnicity and integration understanding population trends and processes, Vol. 3*, (pp. 289–315). Dordrecht: Springer.

Rees, P. (1996). Projecting the national and regional populations of the European Union using migration information. In P. Rees, J. Stillwell, A. Convey, & M. Kupiszewski (Eds.), *Population migration in the European Union* (pp. 331–364). London: Wiley.

Rees, P., Kupiszewski, M., Eyre, H., Wilson, T., & Durham, H. (2001) *The evaluation of regional population projections for the European Union*. Eurostat Working Paper Population and Social Conditions 3/2001/E/n 9. Luxembourg: Office for Official Publications of the European Communities.

Rees, P., Stillwell, J., & Convey, A. (1992). *Intra-community migration and its impact on the demographic structure at the regional level* (Working Paper 92/1). Leeds: School of Geography, University of Leeds.

Rees, P., Wohland, P., Norman, P., & Boden, P. (2010). What happens when international migrants settle? Ethnic population trends and projections for UK local areas. UPTAP Project. Accessed January 22, 2011, from http://www.uptap.net/project36.html

Tromans, N., Natamba, E., Jefferies, J., & Norman, P. (2008). Have national trends in fertility between 1986 and 2006 occurred evenly across England and Wales? *Population Trends*, *133*, 7–19.

Wu, B., Birkin, M., & Rees, P. (2008). A spatial microsimulation model with student agents. *Computers, Environment and Urban Systems*, *32*, 440–453.

Acknowledgements

The School of Geography at the University of Leeds provided funding to enable the symposium to take place from which the contents of this book emerged. The editors are grateful to David Appleyard for redrawing the figures and to each of the authors for their involvement in the project.

Contents

Contributors

Martin Bell School of Geography, Planning and Environmental Management, University of Queensland, QLD, Australia, martin.bell@uq.edu.au

Mark Birkin School of Geography, University of Leeds, Leeds LS2 9JT, UK, m.h.birkin@leeds.ac.uk

Peter Boden School of Geography, University of Leeds, Leeds LS2 9JT, UK, p.boden@leeds.ac.uk

Martin Clarke School of Geography, University of Leeds, Leeds LS2 9JT, UK, m.c.clarke@leeds.ac.uk

Joel Dearden Centre for Advanced Spatial Analysis, University College London, London, UK, joel.dearden@ucl.ac.uk

Adam Dennett Centre for Advanced Spatial Analysis, University College London, Chorley Institute, Pearson Building, London WC1E 7HB, UK, a.dennett@ucl.ac.uk

Dorota Kupiszewska Central European Forum for Migration and Population Research, International Organization for Migration, Twarda 51/55, 00-818 Warszawa, Poland, d.kupisz@twarda.pan.pl

Marek Kupiszewski Institute of Geography and Spatial Organization, Polish Academy of Sciences, Twarda 51/55, 00-818 Warszawa, Poland, m.kupisz@twarda.pan.pl

Les Mayhew Cass Business School, Faculty of Actuarial Science and Insurance, City University, London, UK, lesmayhew@googlemail.com

Salut Muhidin Demography Program at Department of Business, Macquarie University, North Ryde–Sydney, NSW 2113, Australia, Salut.Muhidin@mq.edu.au

Paul Norman School of Geography, University of Leeds, Leeds LS2 9JT, UK, p.d.norman@leeds.ac.uk

Phil Rees School of Geography, University of Leeds, Leeds LS2 9JT, UK, p.h.rees@leeds.ac.uk

John Stillwell School of Geography, University of Leeds, Leeds LS2 9JT, UK, j.c.h.stillwell@leeds.ac.uk

Frans Willekens Netherlands Interdisciplinary Demographic Institute, The Hague, The Netherlands, Willekens@nidi.nl

Alan Wilson Centre for Advanced Spatial Analysis, University College London, London, UK, a.g.wilson@ucl.ac.uk

Tom Wilson School of Geography, Planning and Environmental Management, University of Queensland, QLD, Australia, tom.wilson@uq.edu.au

Chapter 1
The Dynamics of Populations Large and Small: Processes, Models and Futures

Philip Rees

Introduction: Topics and Aims

This chapter is written as a reflection on population changes over the past and future half centuries. Demographic analysis focuses on the processes which add members to populations and which subtract members. The key population processes are therefore fertility which adds babies, mortality which subtracts mainly older adults and migration which adds and subtracts people mainly in the young adult ages. Migration operates on local, national and international scales. Only migration between spatial units alters their population. The chapter looks at both recent trends and the future. Societies that pay attention to their potential futures are normally more successful than societies that pay no attention. As individuals, we are interested in what might happen over our remaining lifetimes, over the remaining lifetimes of our children and grandchildren.

Underpinning the work that I review are a set of methods for analysis of population change, which include the cohort-component model used to carry out population projections, the life table used to compute life expectancies and statistical analyses of various kinds. Multi-state versions of these analyses are now widely used for example to estimate healthy life expectancies (Khoman, Mitchell, & Weale, 2008), to assess the impact of international migration on European populations (Bijak, Kupiszewska, Kupiszewska, Saczuk, & Kicinger, 2007; Bijak, Kupiszewska, & Kupiszewski, 2008) or to evaluate the impact of alternative policy-linked scenarios on European region populations (Rees et al., 2010a). There is a strong push to develop methods to estimate the uncertainties around traditional point values using various statistical methods (Keilman, 2001; Clark, 2003; Wilson & Bell, 2004; Booth, 2004; Alders, Keilman, & Cruijsen, 2007; Keilman, 2007; Lutz, Sanderson, & Scherbov, 2008a). Development of official UK stochastic population projections has started (Shaw, 2008; Rowan & Wright, 2010) but needs further analysis before the status of national statistics is achieved.

P. Rees (✉)
School of Geography, University of Leeds, Leeds LS2 9JT, UK
e-mail: p.h.rees@leeds.ac.uk

J. Stillwell, M. Clarke (eds.), *Population Dynamics and Projection Methods*, Understanding Population Trends and Processes 4, DOI 10.1007/978-90-481-8930-4_1,

The chapter is organized as follows. It begins with an overview of world population growth in the twenty-first century and then breaks this down into growth in the major groupings of countries. Here the dominant theme is the diffusion, at faster and slower paces, of the *first demographic transition* that shifts populations from high to low growth regimes. This world picture informs not only demographic research in individual world regions and countries but is information used by economists for forecasting the long-run potential for economic growth or by climate scientists for assessing future production of greenhouse gases. The chapter then reviews the processes that contribute to population growth and coming changes in that growth starting with fertility. Here we recognize the *second demographic transition* at work which is keeping fertility levels in Europe below replacement level. Then we turn to recent mortality trends and the factors that might alter them, leading to a discussion of alternative futures for life expectancy. Increases in survival to old age and within that life stage are making important contributions to prospective ageing of populations that have already experienced the first and second transitions. A recurring theme in demographic and social research is the direction of travel of inequalities in health, ill-health and mortality, discussed next. Then we turn to the role that international migrant inflows to developed countries play in filling the gaps in labour supply, which in turn change the ethnic composition of those populations in the *third demographic transition*. The consequences of the first and second demographic transitions, together with above replacement fertility in three post Second World War decades, are sustained population ageing that will challenge existing inter-generational arrangements for retirement, health and personal care at older ages. These challenges have been anticipated for decades as a result of demographic analysis but the responses of our political systems have been slow and inadequate. Solutions are being found but there remains the prospect that the divide between privileged insiders and excluded outsiders will become more marked.

World Population Growth

During the twentieth century, as developments in health care and food production have raised life expectancy and improved standards of nutrition, the world's population growth has been greater than at any other time in its history. Although overall population growth rates are estimated by the United Nations (UN) to have peaked in the 1965–1970 period (UN, 2009), the population of the world has more than doubled in size in the intervening years, reaching 6.8 billion in 2009. Population growth is set to continue and is projected to reach 9.1 billion by 2050, at which point it is expected to level off (UN, 2004, 2007). This represents a slowing down of population growth, which at constant rates would be 11 billion by 2050. Much of the population growth of the next 40 years will be concentrated in less developed countries, particularly those of sub-Saharan Africa. The population of the more developed world will remain fairly static, with negative growth rates prevented only by the continuing impact of net immigration from developing countries.

Whilst disparities in population growth remain, declining levels of fertility and continuous improvements in longevity are ageing the world's population at an unprecedented rate. In 2008, there were an estimated 506 million people aged 65 or over, making up 7% of the world's population. By 2040, the 65+ population will number 1.3 billion, 14% of the total. Within 10 years, it is expected that the global population of over 65s will, for the first time, exceed the number of children aged under 5 (UN, 2009).

Figure 1.1 assembles, using some interpolation and estimation, three projection time series of the world's population. In order of simplest to most complex, these are (i) US Census Bureau World Population Trends (US Census Bureau, 2010), (ii) low, medium and high variants of the United Nations World Population Prospects, 2008 Revision (UN, 2009) combined with an earlier Long-Term projection (UN, 2004) and (iii) three percentiles from IIASA's World Population Projections (Lutz et al., 2008a; Lutz, Sanderson, & Scherbov, 2008b). The IIASA projections use a probabilistic methodology (explained later), which creates several hundred or thousand projections by sampling from estimated distributions of leading fertility, mortality and migration drivers. Each sampled path of say total fertility rate (TFR), life expectancy or total net migration generates a new projection. The projection

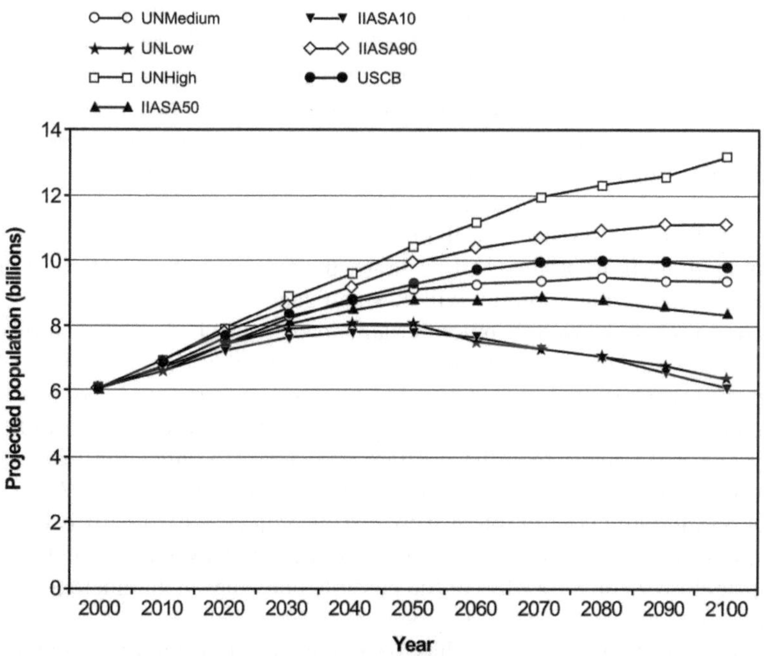

Fig. 1.1 World populations 2000–2100, projected according to the UN, IIASA and USCB (Sources: US Census Bureau, 2010; UN, 2004, 2009; Lutz et al., 2008a, 2008b). (Notes: USCB = US Census Bureau; UN = United Nations; IIASA = International Institute for Applied Systems Analysis)

outcomes are sorted into a ranked database from which percentile distributions of the projected populations can be computed. From these percentile distributions are selected projections at the 10%, 50% and 90% percentiles. When these projections are viewed we state that the chances of the world population being below IIASA10 are only 10%, that the chances of being above the IIASA90 projection are only 10% (100–90%) and there is an 80% chance of the world's population lying between IIASA10 and IIASA90. When variant projections are produced, it is not possible to assign them a position in the probability distribution. There are many technical challenges in preparing such probabilistic projections but as experience is gained they are becoming more common (e.g. Wilson & Bell, 2004; Rowan & Wright, 2010; Keilman, 2001).

All three sources project the same trajectory for the world's population, forecasting that population growth will cease by the end of the century. The contrast between growths in the first and second halves of the century is considerable: between 2000 and 2050 the world population increases by 44% (IIASA50) to 52% (USCB) with 49% for the UN medium projection, whereas between 2050 and 2100 USBC projects a 5% increase, UN projects a 3% increase and IIASA50 projects a 5% decrease.

There are, however, differences between the three projection families. The USCB and UN projections peak at 10.0 and 9.5 billion around 2080 whereas the IIASA50 projection peaks at 8.9 billion around 2070. The IIASA50 projection has lower fertility and mortality assumptions than the USBC and UN projections based on work that links both of these components to rising education levels in developing countries. Increasing participation in education affects women's fertility through postponement during the period of study, through lowering family size goals and through improving receptivity to family planning messages. Mortality is lowered as rising educational attainment translates into more rational approaches to personal health and the better standards of living attained by educated people. If the IIASA projections turn out to be the most accurate, the savings in terms of consumption, resource use and carbon footprint will be significant.

It is important to know how much certainty can be attached to future projections. The UN offers guidance through preparing high and low variant projections. Under the high scenario fertility levels in high fertility countries are kept 0.5 of child above the declining medium projection directory, while under the low scenario the total fertility rate is 0.5 lower than the medium trajectory. The high and low mortality variants reflect different degrees of success in combating the AIDS/HIV epidemic's impacts on mortality. The problem with variant projections is that it is difficult to determine what probability can be assigned to the gap between high and low: does this gap enclose 95 or 67 or 50% of the possible outcomes in an inherently variable process.

The solution to this problem has been the development of probabilistic or stochastic projections, pioneered by Wolfgang Lutz and colleagues at IIASA. The idea is as follows. The variability of each of the components of population change is estimated; a statistical distribution (usually normal) is constructed for leading assumption parameters (e.g. the total fertility rate, the mortality decline rate); the statistical distribution is sampled by generating a random normal deviate which is used to determine a new path for the assumption parameters; a new projection is then run; this procedure is then repeated a large number of times; this results in a distribution of projected populations, which can be described as a cumulative percentile distribution. Key papers describing this set of methods were cited earlier (see Wilson & Bell, 2004, for the clearest account of the operations involved).

Where do the assumption parameter distributions come from? Three methods have been employed: (i) a statistical model (e.g. Auto-Regressive Moving Average or ARIMA model) is fitted to determine the error parameters; (ii) a group of demographic experts is surveyed and asked to estimate key measures associated with projection drivers such as ± one standard deviation from the point projection or the 90/10 percentiles (Shaw, 2008) or (iii) historical errors in the main drivers of population projection are measured (Shaw, 2007). In the UK stochastic projections (Rowan & Wright, 2010), the second and third methods are used and give comparable results. The IIASA world population projections quoted in Fig. 1.1, the latest in a sequence since 1996 (Lutz, Sanderson, & Scherbov, 2004) use expert judgements about the 10/90 percentile range for the component drivers for a set of world regions. In their book on *The End of World Population Growth in the 21st Century* Wolfgang Lutz, Warren Sanderson and Sergei Scherbov say: 'As uncertain as our forecasts are, the likely end of the world's population growth towards the end of the twenty-first century is impossible to overlook' (Lutz et al., 2004, p. 38). In 80% of their hundreds of stochastic projections, population growth has stopped. There is only a 10% chance that the world population in 2100 will exceed 11 billion and only a 10% chance that it will be below its 2000 value.

Differences in Growth Across the World

In Europe, the latest long-term population projections suggest that low birth rates and rising life expectancy combined with continuing net immigration will result in a European Union (EU) population in 2060 that is almost unchanged from today's total but which has a much older age profile (Eurostat, 2008). The median age of the EU population is set to increase from 40 years in 2008 to 48 in 2060, with a mixture of population decline and growth depending upon the vital rate trajectories in individual countries and the influence of net immigration. A 10 year 'window' has been identified during which the countries of the EU have an opportunity to tackle a broad range of policy challenges that are presented by its ageing population. Within this window, the size of the EU labour force is expected to continue to grow before the movement of the baby-boomer cohorts into retirement results in an accelerated process of population ageing with major consequences for economic development and healthcare provision (Commission of the European Communities, 2009).

In 2007, the UK reached a turning point in its own demographic history. For the first time, there were more people over state pension age (SPA) than children aged 0–15 (ONS, 2009). As the large birth cohorts of the 1950s and 1960s move into retirement, the number of adults above SPA as a percentage of the working age

Table 1.1 Rates of population growth, 1950–2050

Major area	1950–1975 (%)	1975–2009 (%)	2009–2050 (Medium) (%)
World	1.89	1.53	0.71
MDR	1.02	0.48	0.08
LDR	2.25	1.82	0.83

Source: UN (2009).

population will continue to increase. Increased life expectancy means an opportunity for us all to live longer, healthier lives but also presents a considerable demographic challenge to retain economic productivity whilst providing healthcare and other services to an increasingly elderly population. The current global economic recession has brought the issue more sharply into focus with the adequacy of pension provision for an ageing population an even greater concern as investments and asset values have plummeted.

The peak growth rate for the world population was in the early 1970s. Since then, falling fertility rates have meant diminishing growth rates. The rates in less developed regions (LDR) are greater than those in more developed regions (MDR) but in 2009–2050, the LDR rates will be lower than the MDR rates were back in 1950–1975 (Table 1.1). This assumes that the fertility transition continues and that LDR total fertility rates fall to replacement or lower. Lutz and Goujon (2001) show how the world's human capital stock will improve in the coming decades to 2030. This improving educational profile will help drive down developing country fertility rates to developed world levels.

The consequence of differences in growth rates between world regions will be shifts in the shares of the world population towards less developed regions, towards Asia, Africa and Latin America and the Caribbean and away from Europe and Northern America in 1950–2009 (Table 1.2).

MDR go from housing one in three of the world's population in 1950 to one in seven in 2050. LDR housed two out three of the world's population in 1950 and will

Table 1.2 Shares of the world population by major regions, 1950–2050

Major area	1950 (%)	2010 (Medium) (%)	2050 (Medium) (%)
More developed regions	32.1	17.9	13.9
Less developed regions	67.9	82.1	86.1
Africa	9.0	15.0	21.8
Asia	55.5	60.3	57.2
Europe	21.6	10.6	7.5
Latin American and Caribbean	6.6	8.5	8.0
Northern America	6.8	5.1	4.9
Oceania	0.5	0.5	0.6

Note: Medium = medium assumption projections.
Source: UN (2009).

house six out of seven in 2050. Africa's share increases by two-thirds between 1950 and 2010 and by 45% from 2010 to 2050. Europe's share shrinks from over one fifth in 1950 to one thirteenth in 2050. Asia's share grows by 5% between 1950 and 2010 but then shrinks back by 3% as East and South East Asia complete the demographic transition to the ageing stage. Latin America with the Caribbean increases its share to 2010 and then falls back. Northern America's share falls continuously through the century but less than does Europe's. Oceania's population is a drop in the world sea but does increase a little by 2050.

Having traced the path of growth of the populations of the world's regions, we now consider what has been happening to the demographic components that contribute to that growth, starting with fertility.

Fertility Trends and the Second Demographic Transition

The broad picture of world population change described above testifies to the diffusion of the demographic transition. Because mortality rates had fallen to low levels in most countries by the end of the twentieth century, the main driver of slowing growth will be continuing falls in fertility in less developed regions. However, this does not mean that growth in those regions will cease. The reason is the demographic momentum built into the current age structure. Demographic momentum is the continued growth of the population despite low fertility rates because of a concentration of the population in the fertile ages (roughly 15–45). As this population ages, growth slows. Demographic momentum means growth will continue for 50–70 years for LDR and some MDR. Even if fertility rates fell to replacement and life expectancies remained static, the population would continue to grow in LDR because the populations are youthful and will have large birth cohorts. In sharp contrast, 45 countries will experience population decreases between 2010 and 2050. These include Germany, Japan, Republic of Korea, Poland, The Russian Federation and Ukraine, all of which will see declines of 10% or more.

Total fertility rates (TFRs) have been below replacement in most European countries for around 40 years. This phenomenon has been labelled *The Second Demographic Transition* by van de Kaa (1987). The continuing below replacement fertility is a function of: women's control of the fertility process through use of the contraceptive pill; the rising cost of raising quality children; rising participation by women in the labour force; the growing participation of women in higher education (where their participation now exceeds that of men); and the rise of individualism and the fall of familism as philosophies influencing the life course. Further downward pressure on fertility rates results from postponement of the age of leaving the parental household (southern Europe) and postponement of age at first birth. Postponement will lead to reduced period fertility, with later recoveries as couples start their families in their 30s. This catch-up has increased TFRs in some countries in the past decade. The cohort measure of completed family size shows less extreme declines (Frejka & Sobotka, 2008).

The consequences of this fertility regime are seen in Fig. 1.2, which plots the period TFRs for NUTS2 regions in Europe in 2005. There is, however, quite a lot of variation in below replacement fertility across the countries of Europe. France, Ireland, Iceland, the United Kingdom, Denmark, Norway, Sweden, Finland, Belgium and the Netherlands have regional fertility rates above 1.5 children per woman. The highest fertility rates are found in France's *Departéments d'Outre Mer* and within 'mainland' Europe, TFRs above 2 are found in France, Finland and Iceland, though only one region, Pohjois-Suomi, experiences above replacement fertility (a TFR of 2.18).

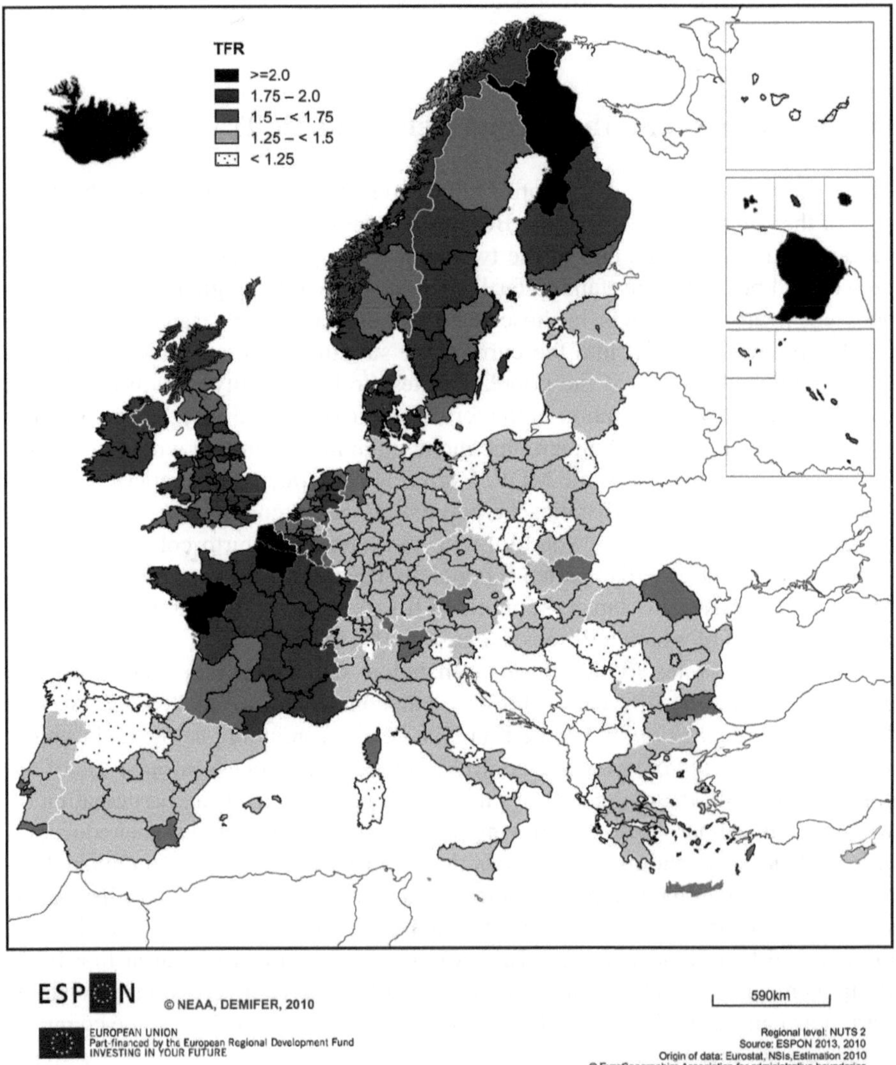

Fig. 1.2 Fertility levels in European regions, 2005 (Source: Rees et al., 2010a)

The other countries of Europe experience TFRs below 1.5 children per woman. Total fertility is below 1.25 in the regions of northern Spain, in some regions of Italy, and in regions in Poland, Slovakia, Hungary, Romania and Bulgaria. In total, 13% of Europe's NUTS2 regions are in this bottom category and 42% in the category of 1.25 TFR up to 1.50, meaning that a majority of regions experienced TFRs below 1.5. Kohler, Billari, and Ortega (2006) provide a full analysis of the causes, implications and policy choices of this phenomenon. They argue that socioeconomic incentives make postponed fertility a rational response to economic uncertainty for young adults and social feedback reinforces this process. Postponement means that family size goals, revealed in surveys still to be two children, are not fulfilled.

These factors play out differently in the countries and regions of higher but still below replacement fertility in western and northern Europe. In these countries, young adults have been able to set up their own households in their 20s, have better child care and support facilities that make maintenance of a mother's work and family careers possible (France and the Nordic countries). Recent history in the higher fertility countries within Europe has been one of rising rates for the native born as a result of catch-up. For example, the TFR in 2001 in the UK was 1.63, but had recovered to 1.94 in 2008. These countries have also experienced immigration of foreign born mothers whose fertility rates are higher than the native population. Recent analysis (Tromans, Natamba, & Jefferies, 2009) suggests rises in both fertility of the UK born and an increase in the size (though not necessarily fertility rates) of foreign born women who made up 14.4% of child-bearing population in 2007 compared with 10.3% in 2001. However, no European national statistical office projects that the increase in period TFRs will continue into the future. It is also not yet clear that really low fertility countries have committed to the family support needed.

American fertility has been at or just below replacement level in recent decades, substantially above fertility rates in Europe. Part of the explanation for these higher rates is the high and increasing share of higher fertility immigrant mothers, particularly from Latin America. But native born Americans also have higher rates which Kohler et al. (2006) suggest stem from more flexible labour markets for women, better private provision of child care, staggered working hours of mothers and fathers and the extended opening hours of retail facilities.

What are the views about future fertility? The UN's view is that total fertility in all countries will converge towards a level of 1.85 children per woman (UN, 2009). High and medium fertility countries are assumed to continue on the downward paths experienced during 1950–2000. Once a TFR of 1.85 is reached, this is adopted as a floor and rates are assumed to be constant thereafter. Rates of fall in fertility are adjusted to match recent evidence. Fertility rates in low-fertility countries are assumed to stay below replacement through to 2050. Where fertility is above 1.85, it is assumed to decline to that floor. Where it is below, it is assumed to increase at 0.05 children every 5 years towards the 1.85 level. This means that the lowest fertility countries will still be well below 1.85 in 2045–2050: 1.2 in 2005–2010 plus eight times 0.05 equals 1.6 in 2045–2050. The evidence of such a rise is not presented: the prevailing view is that lowest fertility countries will continue in the 'low-fertility trap'.

Table 1.3 Future total fertility rate trends

			2045–2050		
Area	1970–1975	2005–2010	Low	Medium	High
UN WPP 2008					
World	4.32	2.56	1.54	2.02	2.51
MDR	2.17	1.64	1.31	1.80	2.30
LDR	5.18	2.73	1.56	2.05	2.53
Europe	2.19	1.50	1.30	1.80	2.29
UK	2.04	1.70	1.35	1.85	2.35
ONS 2008 NPP					
UK	2.04	1.87	1.64	1.84	2.04
			LCL (67%)	Mean	UCL (67%)
Experts (2008)	2.04	1.87	1.54	1.78	2.02

Note: LCL = lower Confidence limit; UCL = upper confidence limit.
Sources: UN (2009), ONS (2009) and Shaw (2008).

Table 1.3 sets out the fertility assumptions from *World Population Prospects 2008* (UN, 2009). The top panel in the table shows the rapid decline in fertility over the past four decades in LDR to 2.73 children in 2005–2010 and to 1.64 children in 2005–2010 in MDR. The UN projected fertility rates fall to 2.02 children by 2050, just below replacement, with a high scenario of 2.51 children and a low of 1.54. The UN sees Europe with the lowest fertility rates of any region, though European demographers would see the TFR of 1.80 in 2050 as optimistic.

The bottom panel sets out the assumptions of the Office for National Statistics in the 2008-based National Population Projections (NPP). The assumptions for the UK are just above the Europe average. Note that the estimate of TFR in 2005–2010 is above the UN forecast. The long-run TFR in the Office for National Statistics (ONS) projections is nearly the same as the UN though the low and high variants are much tighter. The final row of the table tries to develop some idea of the probabilities of low and high outcomes. A panel of six demographic experts was asked to estimate the long-run lower to upper confidence band. If they are right, it suggests that there is only a 15% probability of TFR being at replacement level (around 2.07) or above by 2050 and a 15% probability of being 1.50 or below.

One of the factors mentioned above as influencing fertility trends was the contribution of the immigrant population. What sort of fertility differences do we observe across populations of different origins? Table 1.4 sets out alternative fertility estimates and assumptions used in UK estimates and projections of ethnic group populations. Ethnic minority groups grow out of waves of immigration over time and come to consist of several generations. In the UK, these national origin groups can be traced in the population and indirect estimates of their fertility computed that combine vital statistics, census information and survey data.

The table uses all 16 of the ethnic groups reported in the England and Wales 2001 Census. There are three sets of estimates/assumptions: (i) the ONS estimates developed by Large and Ghosh (2006) using a child-woman ratio method with

Table 1.4 Fertility assumptions used in UK ethnic population estimates and projections

Ethnic group	ONS estimates 2001	OXPOP estimates and assumptions 1996–2006	2006–2011	2031–2036	From 2056	UPTAP estimates and assumptions 2006–2011	From 2021
White							
White: British	1.64	1.72	1.90	1.83	1.83	1.90	1.88
White: Irish	1.51					1.75	1.73
Other White	1.49	1.50	1.68	1.68	1.75	1.71	1.69
Mixed		1.73	1.70	1.72	1.80		
White and Black Caribbean	3.50					1.82	1.78
White and Black African	3.90					2.05	2.01
White and Asian	3.60					1.56	1.53
Other Mixed	3.30					1.62	1.58
Asian							
Indian	1.41	1.64	1.84	1.74	1.70	2.10	1.98
Pakistani	2.12	2.88	2.82	2.30	1.99	2.32	2.12
Bangladeshi	1.94	3.29	2.98	2.29	2.00	2.47	2.29
Other Asian	1.50	1.89	2.02	1.93	1.90	1.74	1.70
Black							
Black Caribbean	1.46	1.89	2.16	2.04	2.00	1.78	1.62
Black African	1.94	2.42	2.34	2.13	1.99	1.82	1.71
Other Black	1.44	2.05	2.42	2.16	2.00	1.74	1.70
Other							
Chinese	1.29	1.31	1.42	1.55	1.70	1.47	1.33
Other Ethnic	1.52	2.03	2.37	2.14	2.00	1.74	1.70
Total	1.63	1.72	1.91	1.86	1.84	1.92	1.93
Data used	2001 Census	Labour Force Survey (Own child method)				Vital statistics, 2001 Census and Labour Force Survey	

Sources: ONS estimates (Large and Ghosh, 2006); OXPOP estimates and assumptions (Coleman & Dubuc, 2010; Coleman, 2010); UPTAP estimates and assumptions (Wohland, Rees, Norman, Boden, & Jasinska, 2010a).

the 2001 Census local authority data; (ii) estimates for use in the Oxford Centre for Population Research (OXPOP) ethnic projection model prepared by Coleman (2010) using pooled Labour Force Survey (LFS) data and the own child method; and (iii) the estimates and assumptions used in the (UPTAP) projections[1] (Wohland

[1] The UPTAP projections are so called because they have been produced through a research project (RES-163-25-0032: 'What happens when international migrants settle? Ethnic group population trends and projections for UK local areas') that was part of the ESRC 'Understanding Population Trends and Processes' initiative.

et al., 2010a). Comparing the ONS and OXPOP estimates for 2001 and 1996–2006 respectively, we see that the ONS child-woman ratio method exaggerates the fertility of mixed groups. The LFS-based estimates are preferable. The ONS estimates are low because 2001 saw the lowest post-1945 TFR. Comparing the OXPOP and UPTAP estimates for 2006–2011, we can see that the use of vital statistics, which ensures that the estimated fertility rates are compatible with observed local authority birth counts, reduces the TFRs for the Asian, Black and Other ethnic groups but increases the Indian TFR. Note that, under the UPTAP estimates, most of the ethnic minority groups have TFRs lower than the White British group, namely the White Irish, Other White, three out of four of the Mixed groups, the Other Asian group, the Black groups and the Other groups. If we compare an average of the OXPOP 2031–2036 and 2056 TFRs with the UPTAP post-2021 assumption, regarding them both as the long-term assumption, the differences are not huge. The main differences are higher OXPOP TFRs for the Black groups (0.35–0.40), the Other Groups (0.30–0.37) and lower TFRs for the Indian group (lower by 0.25). Further analysis is needed to understand how much these variations contribute to the differences in ethnic group projections for the UK as a whole (Norman, Rees, & Wohland, 2010). We now turn to a discussion of the mortality component.

Recent and Future Mortality Trends

Over the last half century, remarkable gains have been made in life expectancy at country, region and world scales (Table 1.5). The top panel of the table reports UN estimates and projections of the life expectancies of both sexes combined. Of course, the gains have not been universal over time and space. Many African countries saw huge retreats in 1990–2010 as a result of the AIDS epidemic, although public health campaigns and the spread of retroviral treatments have stemmed the

Table 1.5 Life expectancies for the world, regions, Europe and the UK, both sexes, 1950–2050

		Estimate		Projection		Change per year		
Unit	Source	1950–1955	1975–1980	2005–2010	2045–2050	1950–1955 to 1975–1980	1975–1980 to 2005–2010	2005–2010 to 2045–2050
World	UN	46.6	60.2	66.4	75.5	0.54	0.21	0.23
MDRs	UN	66.0	72.1	77.1	82.8	0.24	0.17	0.14
LDRs	UN	41.0	57.2	65.6	73.5	0.65	0.28	0.20
Europe	UN	65.6	71.2	75.1	81.5	0.22	0.13	0.16
UK	UN	69.2	72.8	77.2	83.6	0.14	0.15	0.16
UK	ONS		73.9	80.5	86.9		0.23	0.16

Sources: UN (2010) and ONS (2010).

decline in many states. The countries of the Former Soviet Union saw dips in their life expectancies in the decade after the transition and this decline continues for men in Russia. Failed states, such as Afghanistan, Somalia, Haiti or North Korea have not participated in the gains. But in general, mortality trends have been favourable in Asia, Latin America, Northern America and Oceania. EU countries have favourable trends including the new members who have recovered from their post-transition downturn.

The table also compares the UN estimates and projections of life expectancy with alternative estimates that draw on national statistical data. For example, we see that the UN estimates lag behind those of the UK's Office for National Statistics. There is a lag in supplying mortality statistics to the UN by national statistics offices and the UN estimates are conservative. So the ONS estimates the life expectancy of the UK to be 80.5 in 2010 compared with the UN's 77.2 years in 2005–2010, which is lower even allowing for the 3 year difference. The right hand side of the panel reports the average change per year in three periods between 1950 and 2050. In general, the rate of improvement slows down from the earlier period where the gains were made in the control of childhood infectious diseases. Rates of improvement for MDR are lower than those for LDR. Europe's rate of improvement is lower than the rate for MDR in general in the past 30 years, but is projected to be slightly higher. The rates of improvement assumed are relatively conservative and for the UK are lower in 2005–2050 than in 1975–2050. We note that MDR life expectancies for men are catching up with those of women, probably because structural change in employment is moving them out of the more risky occupations (mining, heavy manufacturing, construction and fishing). Women, not being employed in these occupations, do not experience the reward of moving to less risky jobs.

There is an ongoing debate about future mortality trends. On the one hand, there are the *optimists* who say we should extrapolate the gains of the past century into the future because that is what we have seen in the past. On the hand there are *pessimists* who point to new mortality risks that are in the pipeline which will increase disease prevalence and then mortality risk. Oeppen and Vaupel (2002), writing in *Nature*, establish that life expectancies of the top country in world life expectancy rankings over more than a century show a stable linear climb. Consistently, national projections and pensions systems have under-estimated the improvement and so underestimated population ageing and the necessary contribution rate. The National Population Projections Advisory Group (NPPAG) persuaded the ONS to move in their national population projection model (ONS, 2009) from an assumption that used long-term constant mortality rates to one that used continued improvements (a decline in age-specific mortality rates of 1% per annum). In fact, when asked to estimate future life expectancies, the NPPAG was more optimistic (Shaw, 2008) than the ONS projections of life expectancy. Some members argued that the long-term improvement rate should be a decline of 2% per annum in mortality rates, roughly the level of the past 25 years rather than 1% (characteristic of the last 100 years). If you assume a modest but continuing decline of 2% in mortality rates (equivalent to a 0.2 year per year improvement in life expectancy), then after 100 years more than 50% of the population will survive to age 100 for a set of low mortality populations (Canada, Denmark, France, Germany, Italy, Japan, UK, USA) (Christensen, Doblhammer, Rau, & Vaupel, 2009, table 1). Improvement in mortality rates in the period 1992–2005 were even higher than this at 2.7% decline per annum in age-specific mortality rates (Rees et al., 2010a). In designing a very optimistic scenario to show what impact a package of policies could potentially achieve, Rees et al. (2010a) use a decline rate of 3.2% each year, which by 2050 yields a life expectancy of 94 (both sexes) for the population of Europe (31 countries). While such outcomes are not very likely, they do show what might well be achieved by economically, socially and medically advanced societies.

There are contrary views. The pessimists point to episodes of worsening mortality in groups of countries or in particular periods. For example, many countries in Sub-Saharan Africa have seen life expectancies plummet as a result of a continuing HIV/AIDS epidemic. Malaria and TB are still substantial killers and we are running out of effective treatments because of resistance by the parasite or bacterium. Many Former Soviet Union countries (e.g. Russia, Ukraine) have experienced worsening mortality because of increased poverty, rising inequality and persistent alcoholism. Olshansky et al. (2009) point to the potential for obesity trends to translate into life expectancy losses with obese generations reaching old age. They find that if all persons with body mass indexes (BMIs) of 30–35 were to experience the mortality rates of those with a BMI of 24 (when mortality rates are at a minimum), then gains of 0.21–1.08 of a year in life expectancy would be made, depending on race and sex group. The Foresight Panel (Government Office for Science, 2007, p. 32 and figure 18) estimates that if obesity rises as predicted, females will lose around 0.19 of a year and males around 0.37 by 2051. Compared with the gains assumed in the 2008-based NPP of 7.1 years for men and 6.6 years for women, these reductions are small, only 5%.

The Prospective Studies Consortium (2009) writing in *The Lancet* report on the mortality risk experienced by participants in 57 prospective studies in western European and northern America, controlling for age, sex, smoking status and study, against their BMI (kg/m^2) measured at the start. Their evidence comes from about 900,000 adults tracked over many years. The all cause mortality rates are reproduced in Table 1.6.

Mortality rates are high for BMIs below 20 (underweight), which the authors interpret as the effect of weight loss in patients dying of cancer. Mortality rates are lower in the two normal weight ranges (BMI of 20.0–22.5 and 22.5–25.0) and in the lower of overweight categories (25.0–27.5). Then occurs a steep rise in mortality risk which doubles by BMI of 40. When these rates are combined with the projected distributions of BMI in the UK population in 2009 and 2050 from Government Office for Science (2007), they show that the overall mortality rate will increase by 10%. This represents perhaps 10 years of improvement in life expectancy or about 1.5–2 years of the projected gain of 6–7 years in life expectancy. These computations are approximate but suggest that the impact of obesity on life expectancy has been under-estimated. But the obesity epidemic is not going to reverse life

Table 1.6 The relationship between BMI and mortality

	All cause mortality (ages 30–89) annual rates per 1000 population		Percent deaths	
BMI Class	Men	Women	Men	Women
15.0–<17.5	26.4	15.1	0.5	0.3
17.5–<20.0	18.4	10.5	4.1	1.6
20.0–<22.5	15.8	9.5	12.3	4.3
22.5–<25.0	14.5	8.9	19.9	5.5
25.0–<27.5	15.3	9.2	19.8	4.6
27.5–<30.0	16.9	10.4	11.1	3.3
30.0–<32.5	20.5	11.4	5.0	2.3
32.5–<35.0	22.7	13.0	1.7	1.3
35.0–<37.5	26.0	14.7	0.6	0.8
37.5–<40.0	28.2	17.0	0.2	0.4
40.0–<50.0	34.7	19.2	0.1	0.4

Source: Prospective Studies Consortium (2009, figure 2).

expectancy gains in future but instead will slow them down. The main impact will be on the health of those with high BMIs, the costs of their treatment and the lost productivity. We now look at how mortality trends will affect different geographical populations and social groups at different scales.

Inequalities in Health and Mortality

There is a huge literature on inequalities in mortality and health. Here we comment on a few selected recent findings on the UK and US populations. The Marmot Review (2010) provides a comprehensive review of the causes of inequality in health and mortality, stressing the importance of poverty (deprivation of income and access to a good environment) over behaviour (unhealthy life styles), which is seen as context determined.

Inequalities between nations in mortality experience and between regions within a country are a product, in part, of their level of development as indexed by national income per capita. However, this is not the whole story: Richard Wilkinson has accumulated much evidence that there is little linkage between mortality and income above a high threshold and that the degree of socio-economic inequality is a better predictor of a society's level of life expectancy rather than poverty/wealth. This evidence is summarised in the *The Spirit Level* (Wilkinson & Pickett, 2009). The evidence that nations with more inequality have poorer outcomes is stronger than the evidence that regional inequality is more important than deprivation directly. Gravelle and Sutton (2009) have tested the impact of income and relative income on self-assessed health using 19 rounds of the UK General Household Survey (GHS) and find only weak support for the relative deprivation hypothesis with a much stronger effect for deprivation per se.

Successive UK Governments have the goal of reducing health inequalities across population groups and across local populations. One of the means for doing this is to adjust the allocation of National Health Service (NHS) funds distributed to local health authorities (Primary Care Trusts,[2] PCTs) so that more deprived populations get higher allocations (Department of Health, 2010). Resource allocation formulae are devised by the Department of Health; the models change through time as new research is commissioned. The current model is called CARAN, with predecessors AREA and RAWP, and is overseen by a Department of Health Advisory Committee on Resource Allocation (ACRA) (Department of Health, 2008). The main component of CARAN involves predicting health service utilisation, relating populations by age and sex to observed demand. It does not explicitly allow for health inequalities. What did ACRA do about this? After reviewing alternatives, ACRA chose to use disability-free life expectancy as a measure that combined mortality and morbidity for use as a health inequality formula. ACRA debated the soundness of the methodology (prevalence rate of disability and the stationary population variable in a life table). The alternative method was to use a multistate life table model (Rogers, Rogers, & Belanger, 1990) that incorporates transitions between the disabled and able states. The multistate method is conceptually superior but the information on transitions is simply not available for local PCTs. Khoman et al. (2008) compare a multistate method with the prevalence method and found the methods gave similar results as long as the transitions were stable over time. This was a useful result as it would be impossible to measure transition rates between health statuses for PCTs from available surveys such as the Health Survey for England or the British Household Panel Study. The Department of Health measure combined 2001 Census limiting long-term illness rates with local life expectancies for 2006 to create a disability-free life expectancy (DFLE) measure for each PCT, which was differenced from a value slightly above the PCT with the highest DFLE; the bigger the difference measure was for a PCT, the higher was its allocation of NHS funding. The final step needed was a recommendation on the weight to be allocated to the Health Inequalities Index (HEI). ACRA recommended to the Secretary of State for Health a weight of 15%, using a consensus judgement. The Department of Health has commissioned further research on health inequality indexes to build a firmer evidence base for a future HEI and its weighting.

What are the challenges in the UK that ACRA's allocation formula is attempting to meet? How has mortality experience varied across local areas over time and deprivation over the past two decades? Thomas, Dorling, and Davey Smith (2010) have established that a steady increase in inequality in standardized mortality ratios across parliamentary constituencies occurred for the 0–64 ages between 1921–1930 and 1999–2007 and for the 0–74 ages between 1990–2001 and 2006–2007. The

[2]The Conservative-Liberal Democrat Coalition Government has announced in 2010 the intention to remove PCTs as the commissioning authorities within the NHS and create General Practitioner Consortia to become the NHS's commissioning authorities. These new arrangements will be worked out over 2010–2012.

authors report that the differences between highest and lowest local authority life expectancies have increased between 2002 and 2008.

Other studies have shown that geographical inequalities are much higher in the working ages (16–64) than in the childhood ages (0–15) or post-working ages (65+). Most studies ignore differences in mortality experience at the oldest ages. This is becoming a much more important source of inequality as more and more people survive beyond 65 and beyond 75. Figure 1.3 reports on an analysis of local authority life expectancies, which take into account all deaths. Local authorities in the UK are grouped into quintiles by deprivation measured using the Townsend Index in 1991 in the top graph of Fig. 1.3 which shows males and females separately.

The clear association of life expectancy with deprivation is shown in the graphs in the positioning of the quintile time series in order of deprivation. The gaps between worst and best off deprivation quintile are bigger for men than for women. There is an upward trend in life expectancy for all quintiles for both men and women but the gains by men are greater than for women, leading to convergence of the sexes. The gaps between the quintiles fluctuate over time but do expand: the T1–T5 gap expands from 3.2 years in 1991 to 3.6 years in 2006 for men and from 1.9 years to 2.5 years for women. The maximum–minimum range (used in Thomas et al., 2010, figure 2) and the standard deviation of life expectancies for 355 local authority areas all show significant increases in the 1991–2007 period with values of these dispersion indicators all being higher for men than women (Table 1.7). Inequality in mortality outcomes have increased in the UK in the past 2 decades.

However, there is evidence of change in the position of some local authorities between 1991 and 2007, which is shown in the bottom graph of Fig. 1.3. In this graph we classify local authorities into 12 types (using the group classification of Vickers, Rees, & Birkin, 2003). As with the deprivation quintiles, the groups of local authorities move upwards reflecting national improvements. The Commuter Belt, Prosperous Urbanites and Rural Britain types have the highest life expectancies; Established Urban Centres, Industrial Legacy and Cosmopolitan Inner London types have the lowest life expectancies. The gap between the upper and lower trios expands over the period. Established Urban Centres lag behind and Industrial Legacy local authorities have a worsening position. But the most remarkable change occurs in the Mercantile Inner London group of authorities (Camden, City of London, Hammersmith and Fulham, Islington, Kensington and Chelsea, Wandsworth and Westminster): at the start of the period this group has the second lowest life expectancy; by the end of the period it has the third highest. This upward mobility in the life expectancy rankings reflects the upward social mobility of its population and in-migration of white collar professionals and business people to take up jobs in London's expanding financial and national government sectors. One might term this climb of a group of local authorities as the 'Hoffenheim' effect after the German Bundesliga small town team that rose to top of the football league in the 2008–2009 season (Wikipedia, 2010a).

Evidence from the United States shows what can happen to life expectancy in countries with high and increasing levels of inequality. Ezzati, Friedman, Kulkarni, and Murray (2008), writing in the *Public Library of Science*, measure

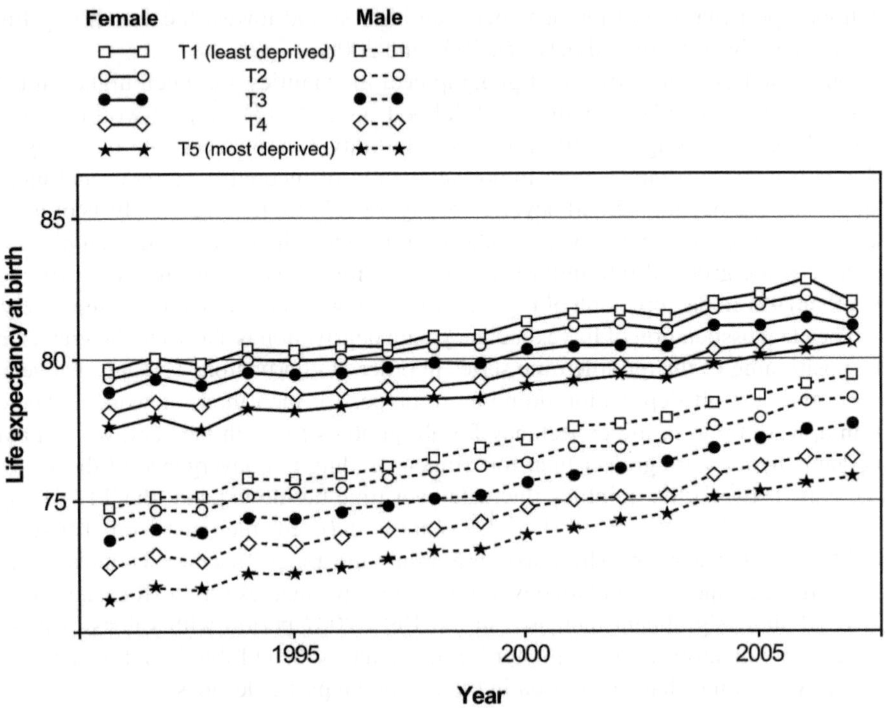

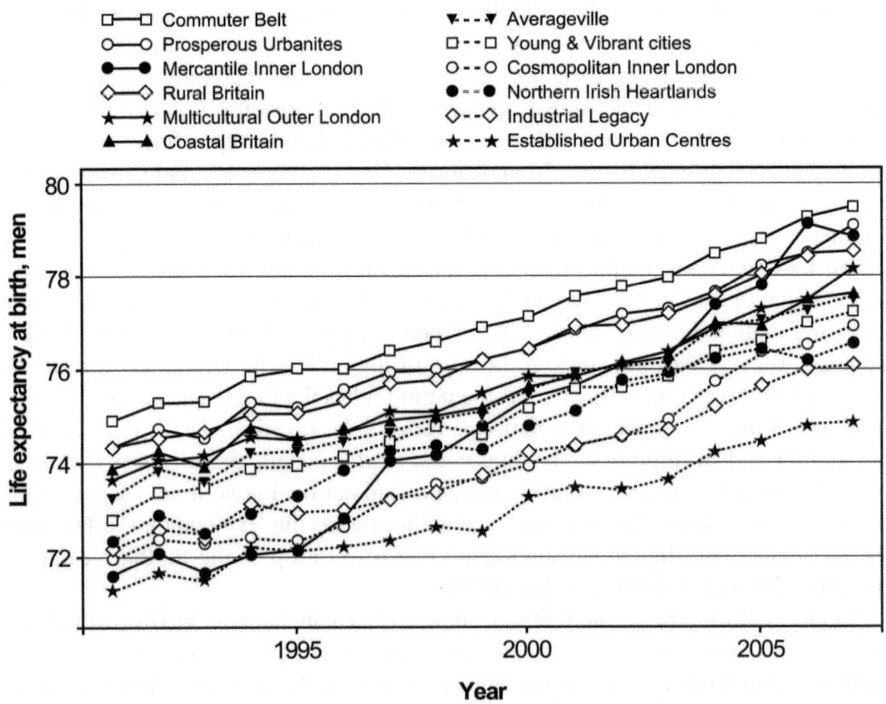

Fig. 1.3 Life expectancies for UK local authorities classified by deprivation quintile, males and females, and by local authority type, males, 1981–2007 (Source: Wohland, Rees, & Norman, 2009)

Table 1.7 Changes in the range and standard deviation of life expectancies in local authorities, UK, 1991–2007

	Men		Women	
Statistic	1991	2007	1991	2007
Max–min range	9.78	13.32	6.82	9.06
Standard deviation	1.55	1.88	1.25	1.36

life expectancy change between 1983 and 1999 for US counties. Table 1.8 sets out a summary of their results. In 1961–1983, 97% of US counties experienced life expectancy gains for men and 99% for women. In the following period, 1983–1999, 13% of counties posted lower life expectancies for men and 36% of counties experienced this for women. The US is one of the richest countries in the world and spends the highest percentage of its gross domestic product on health. However, its health care system was clearly dysfunctional in delivering these outcomes. Of concern is a remark by the authors of the study that 'We used mortality statistics from the National Center for Health Statistics [NCHS] . . . between 1961 and 1999. Data for analyses in subsequent years were not provided to us by the NCHS'. A society needs to be able to monitor its health status in order to put in place improvement policies. The Health Care legislation pushed through the US Congress by the Obama Administration is a recognition that collective action has to be taken (Wikipedia, 2010b).

The geographical variation of mortality reflects, in the main, the variation in the income and social well-being of the populations of the geographical units. Mortality experience also differs across other dimensions of difference such as ethnicity. Differences in mortality experience between ethnic groups are difficult to measure when ethnicity is not recorded on the death certificate. This is the case in the UK (though country of birth is recorded) and ethnic mortality differences were not incorporated into official ethnic population estimates (Large & Ghosh, 2006) or into

Table 1.8 Life expectancy changes, US counties, 1961–1983 and 1983–1999

		Percent distribution of counties			
Change	Significance (90% level)	1961–1983 Men	1961–1983 Women	1983–1999 Men	1983–1999 Women
Increase	Significantly more than national mean	19	21	9	9
Increase	Not significantly different from national mean	63	64	59	23
Increase	Significantly less than mean	14	14	17	1
Zero	Not significantly different from national mean	0	0	0	31
Decline	Not significant	3	1	13	30
Decline	Significant	0	0	0	6
	All US	100	100	100	100

Source: Computed from Ezzati et al. (2008, table 2).

Table 1.9 Estimates of the life expectancies of ethnic groups in the UK, 2001

Ethnic group	Life expectancy	Ethnic group	Life expectancy
Chinese	82.1	Other Asian	79.5
Other Ethnic	81.5	White-Black African	79.5
Other White	81.3	Indian	79.3
White British	80.5	Black Caribbean	79.1
All groups	80.5	White-Black Caribbean	78.7
Black African	80.4	Other Black	78.5
White-Irish	80.3	Bangladeshi	77.7
White-Asian	80.0	Pakistani	77.3
Other Mixed	79.9		

ethnic group projections (Coleman & Scherbov, 2005; Rees & Parsons, 2006). Rees and Wohland (2008) and Rees, Wohland, and Norman (2009) constructed estimates of life tables for the UK's ethnic group using a combination of proxy variables (standardised limiting long-term illness), local mortality statistics and some models. The estimates were aligned with local authority life tables. Summary results are given in Table 1.9 for the 16 ethnic groups used in the 2001 Census. The spatial patterns of life expectancy for the 16 ethnic groups are mapped in Rees and Wohland (2008), Rees et al. (2009) and Wohland and Rees (2010).

Four groups lie above or at the whole population mean and 12 groups lie below it. The range between lowest and highest groups is only 4.8 years. The differences between the groups are moderate when compared with the local authority differences discussed earlier and in comparison with the few other countries where such estimates are available (e.g. US or New Zealand). Contrast the UK experience with that in the US set out in Table 1.10. Murray et al. (2006) construct life tables for race and county combinations which pick out the poorest in American society. Black men in high-risk urban areas have life expectancies 16 years lower than Asians in America in 1982 and 2001 and for women the difference is 15 years in 1982 and 13 years in 2001. The authors write as follows: 'Disparities in mortality across the eight Americas, each consisting of millions or tens of millions of Americans, are enormous by all international standards. ... Because policies aimed at reducing fundamental socioeconomic inequalities are currently practically absent in the US, health disparities will have to be at least partly addressed through public health strategies that reduce risk factors for chronic diseases and injuries' (Murray et al., 2006, p. 1513).

The Third Demographic Transition: Changing Ethnic Composition

Coleman (2004, 2006) has suggested many developed nations have moved on into a *third demographic transition*. This involves the following processes: population ageing as a result of continuing low fertility and increasing survival to and within

Table 1.10 Life expectancies in the 'Eight Americas', 1982 and 2001

The 8 Americas: race-county groups	Population 2000	Men			Women		
		1982	2001	Diff	1981	2001	Diff
Asian	10.4	79	82	3	89	88	−1
Northland low-income rural white	3.6	73	76	3	80	82	2
Middle America	214	72	75	3	78	80	2
Low-income whites in Appalachia and the Mississippi Valley	16.6	70	72	2	79	78	−1
Western Native America	1.0	65	70	5	74	76	2
Black Middle America	23.4	65	70	5	74	76	2
Southern low-income and rural black	5.8	64	67	2	74	75	1
High-risk urban black	7.5	63	66	3	74	75	1
Total	282.3	71.1	74.3	3.1	78.0	79.6	1.7
US life expectancy		70.8	74.4	3.6	78.1	79.8	1.7

Note: Pop = Population 2000 Census in millions.
Source: Murray et al. (2006, figure 3). US life expectancy: World Bank (2010).

older ages; a labour supply shortage (that will be increased as a result of the retirement of the baby boomers over the next 30 years); and (net) immigration from more youthful countries, which results in changes to the ethnic composition of the population. The process is driven by international migration from poorer but youthful countries to richer but older countries but would continue, even if net immigration were reduced to zero because of the demographic momentum built into a youthful immigrant origin population.

Most national and international projections neglect this composition effect. Most European projections use foreign birth or foreign nationality as the indicator for the 'different' population (see Coleman, 2006 for a review). After the first or occasionally the second generation, the foreign origin population disappears from view. Since the 1980s, the UK national statistics agency and social scientists have adopted a different perspective: such communities of immigrant-origin will persist through many generations and should be allowed to self-identify in official surveys and censuses. From the 1991 Census it has been possible to make estimates of ethnic group populations and to carry out ethnic group projections (see Wohland, Rees, Norman, & Boden, 2010b; Rees, Wohland, Norman, & Boden, 2010b for reviews). Table 1.11 reports on the results of a recent set of ethnic group population projections for local areas in the UK (Wohland et al., 2010a).

For convenience, we have aggregated the projected populations from 355 areas (352 local authorities in England plus Wales, Scotland and Northern Ireland) into one UK figure and from 16 ethnic groups to five broad groups (White, Mixed, Asian, Black and Other). The table reports on three projections. The first, called BENCH, is a benchmark projection which adopts the rates, intensities and flows for ethnic group

Table 1.11 Projected populations for five ethnic groups, UK, 2001–2051

		BENCH		TREND		UPTAP	
Ethnic group	2001	2011	2051	2011	2051	2011	2051
Populations (1000s)							
White	54,384	54,838	51,508	57,087	65,239	56,835	60,996
Mixed	687	976	1,999	1,029	2,572	1,013	1,945
Asian	2,373	3,031	5,244	3,133	6,251	3,074	4,940
Black	1,174	1,442	2,094	1,492	2,523	1,456	1,868
Other	492	748	1,385	790	1,664	729	956
UK Total	59,111	61,035	62,230	63,531	78,249	63,108	70,705
Time series (2001=100)							
White	100	101	95	105	120	105	112
Mixed	100	142	291	150	374	147	283
Asian	100	128	221	132	263	130	208
Black	100	123	178	127	215	124	159
Other	100	152	282	161	338	148	194
UK Total	100	103	105	107	132	107	120
Percentage shares							
White	92.0	89.8	82.8	89.9	83.4	90.1	86.3
Mixed	1.2	1.6	3.2	1.6	3.3	1.6	2.8
Asian	4.0	5.0	8.4	4.9	8.0	4.9	7.0
Black	2.0	2.4	3.4	2.3	3.2	2.3	2.6
Other	0.8	1.2	2.2	1.2	2.1	1.2	1.4
UK Total	100.0	100.0	100.0	100.0	100.0	100.0	100.0

Notes: BENCH = Benchmark projection using emigration flows; TREND = Projection aligned to ONS NPP assumptions; UPTAP = Projection with revised assumptions and emigration rates.
Source: Wohland et al. (2010a).

in UK local areas for 1 year periods around the 2001 Census and then assumes these intensities apply into the future. The second, called TREND, uses information for the intensities for 2001–2007 drawing on official estimates and other sources and adopts assumptions for the future beyond 2007 that are aligned with the 2008-based national population projections (NPP) for the UK as a whole. The third projection, called UPTAP, revises the TREND assumptions a little and adopts a different model for emigration, projecting it not as a flow number (the NPP procedure for net international migration) but as an emigration rate multiplied by a population at risk (of each ethnic group in each local area), while assuming future immigration will remain a constant flow.

Population projections are always prisoners of their gestation environment. They are heavily influenced by the demographic rates and flows of the immediate past, even though considerable cohort analysis underpins the projection assumptions. So the BENCH projection reflects the demographic regime around 2001, in which the TFR was lower than in any year in the previous half century, when mortality was high compared with a decade later and when net immigration was much lower than in subsequent years. The result is a projected UK population of only 62 millions in

2051 on a declining trajectory, having peaked in the 2030s. The TREND scenario projects populations of 78.2 million which is slightly higher than the 77.1 million of the 2008-based NPP, the difference being attributable to the larger number of groups used in the former. When we switch the international migration model to one which uses emigration rates and UK populations at risk, then the projected UK population is only 70.7 million. The arguments for and against this model switch are rehearsed in Rees et al. (2010b).

The different ethnic groups grow at very different rates. The second panel of Table 1.11 sets out time series indicators for each group where a figure of 100 represents the group population in 2001. Under the BENCH projection, the White population shrinks while under the other two projections it grows a little (20 and 12%). The Mixed group, very youthful in 2001, and receiving birth contributions from other groups is the fastest growing, albeit from a small base, nearly doubling in the BENCH and UPTAP projections and increasing by 274% in the TREND projection. This shows the power of demographic momentum as a driver of growth as this group receives only a small contribution from international migration. The Asian group grows strongly followed by the Other group and Black group. For these latter groups, using an emigration rates model reduces their population growth compared with the BENCH and TREND models. The Other group contains Chinese immigrants who come to study, many of whom will return to China and Hong Kong on completion of study. The Other group also contains the Other ethnics sub-group in which most asylum seekers and refugees fall, who again will return to their origins, if these are deemed safe.

The result of these differences in growth is inevitable change in ethnic composition. In the UPTAP projections, the White population shrinks from 92 to 86.3% of the total. The Mixed groups expand from 1.2 in 2001 to 1.6 in 2011 to 2.8% in 2051. Asians rise from 4 to 4.9% to 7% in 2051. The Black group expands from 2 to 2.3% to 2.6% of the population. The Other group's share rises from 0.8 to 1.2% to 1.4%. The UK population will be more diverse in 2051 than it was in 2001. Further details of these projections are discussed in Wohland et al. (2010a, 2010b) and Rees et al. (2010b).

Population Ageing

Once populations have completed the demographic transition, then they will undergo population ageing, which involves an increasing mean age, a larger number of older people, a rise in the share of the population that is old. As a consequence of ageing, the ratios between child ages, working ages and older ages change. At first the working age population increases, the child dependency ratio falls and the older age ratio rises. This process boosts productivity of the national population because of an increasing concentration in the working ages. Later, as larger birth cohorts reach old age, this favourable situation recedes and old age dependency ratios rise sharply. Unless countervailing processes such as longer working lives or baby booms are at work, national productivity will fall and financial and service support for the older population will become more difficult.

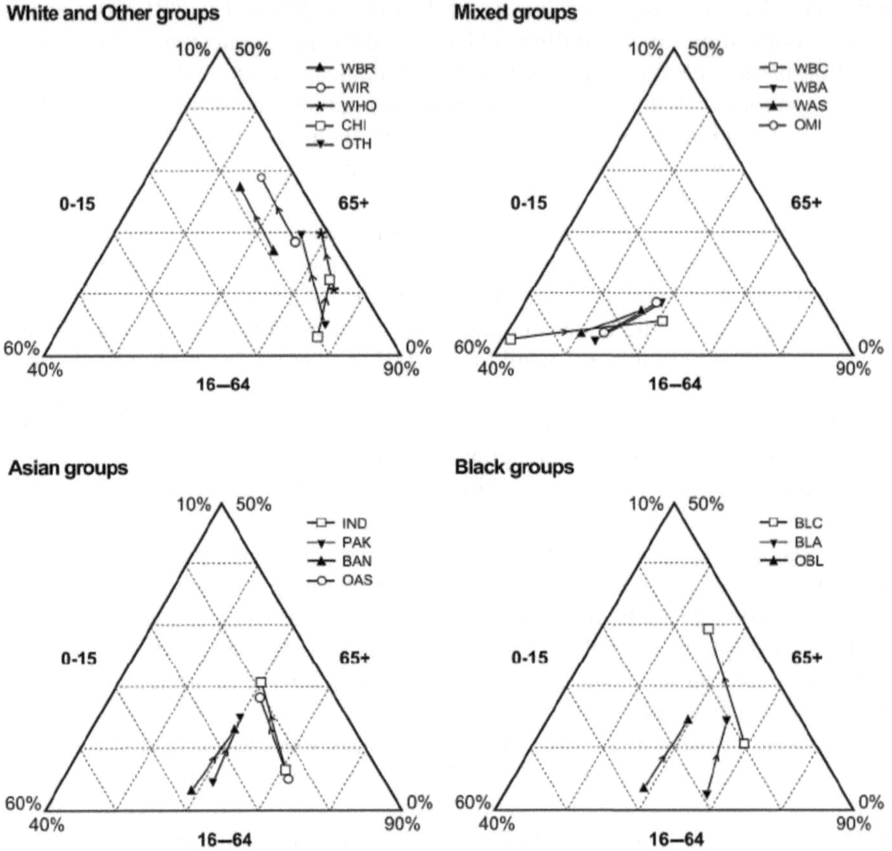

Fig. 1.4 Changes in the age structure of ethnic group populations, UPTAP projection, 2001–2051 (Source: Wohland et al., 2010a)

The ageing process also applies to sub-population such as ethnic groups. Figure 1.4 captures the ageing process in 16 ethnic groups in the UK population, drawing on the projection results discussed in the previous section. The ethnic groups are arranged in four separate graphs for ease of reading. The arrows connect the 2001 position of a group in the graph with its position in 2051. Groups move around the triangular space on a particular path. The youngest groups (Mixed) (top right hand graph) are situated close to bottom left-hand corner and move rightward, increasing their working age share but not yet their older population share. Then there are a set of groups (e.g. Black African in the bottom right graph) that start about half way across the graph close to the bottom that move in a north-east direction keeping their share of the working age population stable but reducing the child share of the population and increasing the elder share. Then there are groups (Indian, Other Asian in the bottom left hand graph) positioned with high percentages in the working ages, low percentages in the older ages which move in a 'northerly' direction increasing their elder shares while seeing their working age shares decrease. Finally, there is a set of groups (e.g. White British, White Irish) which already have a high percentage in the older group and which see this percentage increase as the labour force and child ages decrease. These age structure changes have important implications for how societies function. Boden and Rees (2010) provide a useful overview in relation to demographic developments in the north of England.

Concluding Remarks

We have voyaged on a tour of the dynamics of populations large and small, which I hope you have found of interest. We have observed alternative trajectories of world population growth to the end of this century and exposed the variations between major regions. We have explored different assumptions about future fertility and mortality globally and for different ethnic groups in the UK. We have demonstrated the extent of change in the ethnic population of the UK and the exposed how the ageing process will impact on the age structure of ethnic group populations. There is a lot we do not know and much we need to confirm or to discover. Improving our population statistics is a vital ingredient on the way to such knowledge. Which statistics are needed is informed by our ongoing research into population issues and associated methods. Demographic analysis is enormously important for society so that we understand what population changes are taking place and will unfold and contribute to sensible policy formulation. Demography gives us time but we need to use it well.

References

Alders, M., Keilman, N., & Cruijsen, H. (2007). Assumptions for long-term stochastic population forecasts in 18 European countries. *European Journal of Population, 23*(1), 33–69.

Bijak, J., Kupiszewska, D., & Kupiszewski, M. (2008). Replacement migration revisited: Simulations of the effects of selected population and labour market strategies for the ageing Europe, 2002–2052. *Population Research and Policy Review, 27*, 321–342.

Bijak, J., Kupiszewska, D., Kupiszewski, M., Saczuk, K., & Kicinger, A. (2007). Population and labour force projections for 27 European countries, 2002–2052: Impact of international migration on population ageing. *European Journal of Population, 23*, 1–31.

Boden, P., & Rees, P. (2010). *Demographic change and its impact upon the Northern Way regions*. Paper presented at the Symposium on Economic Development in the Northern England, The Lowry Centre, Salford. Online at http://www.thenorthernway.co.uk/page.asp?id=822

Booth, H. (2004). On the importance of being uncertain: Forecasting population futures for Australia. *People and Place, 12*(2), 1–12.

Christensen, K., Doblhammer, G., Rau, R., & Vaupel, J. W. (2009). Ageing populations: The challenges ahead. *Lancet, 373*, 1196–1208.

Clark, J. S. (2003). Uncertainty and variability in demography and population growth: A hierarchical approach. *Ecology, 84*(6), 1370–1381.

Coleman, D. (2004, September 13). *Migration in the 21st century: A third demographic transition in the making?* Plenary Address to the British Society for Population Studies annual conference, Leicester.

Coleman, D. (2006). Immigration and ethnic change in low-fertility countries: A third demographic transition. *Population and Development Review, 32*(3), 401–446.

Coleman, D. (2010). Projections of the ethnic minority populations of the United Kingdom 2006–2056. *Population and Development Review, 36*(3), 441–486.

Coleman, D. A., & Dubuc, S. (2010). The fertility of ethnic minorities in the UK, 1960s–2006. *Population Studies, 64*(1), 19–41. doi: 1080/00324720903391201.

Coleman, D., & Scherbov, S. (2005). *Immigration and ethnic change in low-fertility countries – Towards a new demographic transition?* Paper presented at the Population Association of America annual meeting, Philadelphia. Online at http://www.apsoc.ox.ac.uk/Oxpop/publications%20files/WP29.pdf

Commission of the European Communities. (2009). *Dealing with the impact of an ageing population in the EU (2009 Ageing Report).* Communication from the Commission to the European Parliament, the Council, the European Economic and Social Committee and the Committee of the Regions, COM (2009) 180/4. Online at http://ec.europa.eu/economy_finance/publications/publication14996_en.pdf

Department of Health. (2008). *Report of the Advisory Committee on Resource allocation.* Online at http://www.dh.gov.uk/en/Publicationsandstatistics/Publications/PublicationsPolicyAndGuidance/DH_091484

Department of Health. (2010). *NHS allocations.* Online at http://webarchive.nationalarchives.gov.uk/+/www.dh.gov.uk/en/Managingyourorganisation/Financeandplanning/Allocations/index.htm

Ezzati, M., Friedman, A. B., Kulkarni, S. C., & Murray, C. J. L. (2008). The reversal of fortunes: Trends in county mortality and cross-county mortality disparities in the United States. *PLoS Medicine, 5*(4), e66. 1–12.

Eurostat. (2008). Ageing characterises the demographic perspectives of the European societies. *Eurostat Statistics in Focus 72/2008.* Online at http://epp.eurostat.ec.europa.eu/cache/ITY_OFFPUB/KS-SF-08-072/EN/KS-SF-08-072-EN.PDF

Frejka, T., & Sobotka, T. (2008). Overview Chapter 1 : Fertility in Europe: Diverse, delayed and below replacement. *Demographic Research, 19*(3), 15–46.

Government Office for Science. (2007). *Foresight. Tackling obesities: Future choices – Modelling future trends in obesity and their impact on health* (2nd ed.). Online at http://www.foresight.gov.uk/Obesity/14.pdf

Gravelle, H., & Sutton, M. (2009). Income, relative income, and self-reported health in Britain 1979–2000. *Health Economics, 18*, 125–145. doi: 10.1002/hec.1354.

Keilman, N. (2001, August 2). Uncertain population forecasts. *Nature, 412*, 490–491.

Keilman, N. (2007). UK national population projections in perspective: How successful compared to those in other European countries? *Population Trends, 129*, 20–30.

Khoman, E., Mitchell, J., & Weale, M. (2008). Incidence-based estimates of life expectancy of the healthy for the UK: Coherence between transition probabilities and aggregate life tables. *Journal of the Royal Statistical Society, A, 171*(1), 203–222.

Kohler, H-P., Billari, F. C., & Ortega, J. A. (2006). Low rertility in Europe: Causes, implications and policy options. In F. R. Harris (Ed.), *The baby bust: Who will do the work? Who will pay the taxes?*(pp. 48–109). Lanham, MD: Rowman & Littlefield.

Large, P., & Ghosh, K. (2006). A methodology for estimating the population by ethnic group for areas within England. *Population Trends, 123*, 21–31.

Lutz, W., & Goujon, A. (2001). The world's changing human capital stock: Multi-state population projections by educational attainment. *Population and Development Review, 27*(2), 323–339.

Lutz, W., Sanderson, W., & Scherbov, S. (Eds.). (2004). *The end of world population growth in the 21st century*(2nd ed.). London: Earthscan Books.

Lutz, W., Sanderson, W., & Scherbov, S. (2008a, January 20). The coming acceleration of global population ageing. *Nature, 451*, 716–719. doi: 10.1038/nature06516. Letter. Reproduced with supplementary information in POPNET, Population Newsletter, No. 39, Winter 2007/8, International Institute for Applied Systems Analysis. Online at http://www.iiasa.ac.at/Research/POP/POPNET/index.html

Lutz, W., Sanderson, W., & Scherbov, S. (2008b). *IIASA's 2007 Probabilistic world population projections*. IIASA world population program online data base of results 2008, http://www.iiasa.ac.at/Research/POP/proj07/index.html?sb=5

Marmot Review. (2010). *Fair society, healthy lives*. Strategic review of health inequalities in England post-2010. Online at www.ucl.ac.uk/marmotreview

Murray, C. J. L., Kulkarni, S. C., Michaud, C., Tomijima, N., Bulzacchelli, M. T., Iandiorio, T. J., et al. (2006). Eight Americas: Investigating mortality disparities across races, counties, and race-counties in the United States. *Public Library of Science Medicine, 3*(9), e260.

Norman, P., Rees, P., & Wohland, P. (2010). *The estimation and application of ethnic group fertility rates in a projection of sub-national populations in the UK*. Draft paper, School of Geography, University of Leeds, Leeds.

Oeppen, J., & Vaupel, J. W. (2002). Broken limits to life expectancy. *Science, 296*, 1029–1031.

Olshansky, S. J., Passaro, D. J., Hershow, R. C., Layden, J., Carnes, B. A., Brody, J., et al. (2009). A potential decline in life expectancy in the United States in the 21st century. *New England Journal of Medicine, 352*(11), 1138–1145.

ONS. (2009). *2008-based national population projections*. Online at http://www.statistics.gov.uk/downloads/theme_population/NPP2008/NatPopProj2008.pdf

ONS. (2010). *Latest 2008-based period and cohort life expectancy tables*. File period_cohort_tables_index08[1].pdf. Online at http://www.statistics.gov.uk/statbase/Product.asp?vlnk=15098

Prospective Studies Consortium. (2009). Body-mass index and cause-specific mortality in 900 000 adults: Collaborative analyses of 57 prospective studies. *The Lancet, 373*(9669), 1083–1096.

Rees, P., Boden, P., Dennett, A., Stillwell, J., de Jong, A., & ter Veer, M. (2010a). Deliverable 6: Report on scenarios and a database of scenario drivers. The ESPON 2013 Programme, DEMIFER, Demographic and migratory flows affecting European regions and cities. *Applied Research Project* 2013/1/3.

Rees, P., & Parsons, J. (2006). *Socio-demographic scenarios for children to 2020.*York: Joseph Rowntree Foundation.

Rees, P., & Wohland, P. (2008). *Estimates of ethnic mortality in the UK* (Working paper 08/04). Leeds: School of Geography, University of Leeds. Online at http://www.geog.leeds.ac.uk/wpapers/08-04.pdf

Rees, P., Wohland, P., & Norman, P. (2009). The estimation of mortality for ethnic groups at local scale within the united kingdom. *Social Science and Medicine*, 69, 1592–1607

Rees, P., Wohland, P., Norman, P., & Boden, P. (2010b, July). Ethnic population projections for the UK, 2001–2051. *Journal of Population Research* (submitted).

Rogers, A., Rogers, R., & Belanger, A. (1990). Longer life but worse health? Measurement and dynamics. *The Gerontologist, 30*(5), 640–649.

Rowan, S., & Wright, E. (2010, April 28–30). *Developing stochastic population forecasts for the United Kingdom: Progress report and plans for future work.* Item 13 – Stochastic national demographics projections, Conference of European Statisticians, Statistical Office of the European Union (Eurostat), Joint Eurostat/UNECE Work Session on Demographic Projections, Lisbon, Portugal. Online at www.unece.org/stats/documents/ece/ces/ge.11/2010/wp.40.e.pdf

Shaw, C. (2007). Fifty years of United Kingdom national population projections: How accurate have they been? *Population Trends, 128L*, 8–23.

Shaw, C. (2008). The national population projections advisory group: Results from a questionnaire about future trends in fertility, mortality and migration. *Population Trends, 134*, 42–53.

Thomas, B., Dorling, D., & Davey Smith, G. (2010, July 23). Inequalities in premature mortality in Britain: Observational study from 1921 to 2007. *British Medical Journal, 341*, c3639. doi: 10.1136/bmj.c3639 (22 July 2010).

Tromans, N., Natamba, E., & Jefferies, J. (2009). Have women born outside the UK driven the rise in UK births since 2001? *Population Trends, 136*, 28–42.

UN. (2004). *World population to 2300*. New York: United Nations.

UN. (2007). *World population prospects: The 2006 revision* (Working Paper No. ESA/P/WP.202). Department of Economic and Social Affairs Population Division. Online at http://www.un.org/esa/population/publications/wpp2006/WPP2006_Highlights_rev.pdf

UN. (2009). *World population prospects. The 2008 revision* (Vol. 1). Comprehensive tables, Department of Economic and Social Affairs, Population Division, ST/ESA/SER.A/287. New York: United Nations. Online at http://esa.un.org/unpd/wpp2008/peps_documents.htm

UN. (2010). *World population prospects. The 2008 revision*. Population Database. Online at http://esa.un.org/UNPP/

US Census Bureau. (2010). *International data base (IDB)*. World population summary. Online at http://www.census.gov/ipc/www/idb/worldpopinfo.php

van de Kaa, D. (1987). Europe's second demographic transition. *Population Bulletin, 42*(1), 64.

Vickers, D., Rees, P., & Birkin, M. (2003). *New classification of UK local authorities using 2001 census key statistics* (Working Paper 03/3). Leeds: School of Geography, University of Leeds. Online at http://www.geog.leeds.ac.uk/wpapers/03-3.pdf

Wikipedia. (2010a). *TSG 1899 Hoffenheim*. Online at http://en.wikipedia.org/wiki/TSG_1899_Hoffenheim

Wikipedia. (2010b). *Health care reform in the United States*. Online at http://en.wikipedia.org/wiki/Health_care_reform_in_the_United_States

Wilkinson, R., & Pickett, K. (2009). *The spirit level: Why more equal societies almost always do better.*London: Allen Lane.

Wilson, T., & Bell, M. (2004). Australia's uncertain demographic future. *Demographic Research, 11*(8), 195–234.

Wohland, P., Rees, P., & Norman, P. (2009, August 6). *Trends in life expectancy in the UK: How have inequalities changed for local areas in the UK and what can we expect for the future?* Paper presentation at the 5th Biennial Population Geographies conference, Dartmouth College, New Hampshire. Online at http://www.geog.leeds.ac.uk/research/projects/migrants/presentations.html

Wohland, P., & Rees, P. (2010). Life expectancy variation across England's local areas by ethnic group in 2001. *Journal of Maps, v2010*, 354–359.

Wohland, P., Rees, P., Norman, P., Boden, P., & Jasinska, M. (2010a). *Ethnic population projections for the UK and local areas, 2001–2051* (Working Paper 10/02). Leeds: School of Geography, University of Leeds. Online at http://www.geog.leeds.ac.uk/research/wpapers

Wohland, P., Rees, P., Norman, P., & Boden, P. (2010b, August). Alternative futures for the UK's ethnic populations: Sensitivity and scenario analyses. *Journal of Population Research* (in press).

World Bank. (2010). *Life expectancy at birth, total (years)*. Databank. http://data.worldbank.org/indicator/SP.DYN.LE00.IN

Chapter 2
Population Accounts

Frans Willekens

Introduction

Phil Rees pioneered multistate demographic accounts. A population account is a two- or multi-dimensional table that integrates data on population change in a systematic and consistent way. It integrates information on population stocks and flows, demographic events such as births, deaths and migration that cause population change. In a set of population accounts, every person and every relevant demographic event is accounted for. Missing data are estimated from available information, rules of accounting (e.g. no double count) and necessary assumptions. The outcome is a table with consistent data on population size, structure and change. Accounting equations link stocks and flows. Accounts become particularly useful when they serve as the basis for demographic modelling. Account-based models estimate the parameters of the model from the population account.

In their book entitled *Spatial Population Analysis*, Phil Rees and Alan Wilson (1977) showed that population accounting is the keystone of demographic modelling. Many modelling problems that previously could be solved on an ad hoc basis only can be solved comprehensively and more effectively if an accounting framework is adopted. That is particularly so when data are incomplete, which most often is the case. Suppose we need information on migration flows by region of origin and region of destination but the available data is limited to (i) arrivals and departures by region and (ii) a selection of migration flows. To obtain an internally consistent set of migration figures, all flows must be determined simultaneously.

In this chapter, the main principles of demographic accounting are reviewed and it is demonstrated that the approach proposed by Rees and Wilson in 1977 is an Expectation-Maximization (EM) algorithm *avant-la-lettre*. What this means is that the accounting method designed by Rees and Wilson incorporates the core features of the EM algorithm, which was not yet available at that time and a dominant method of statistics only later. The EM algorithm is a generic method for estimating model

F. Willekens (✉)
Netherlands Interdisciplinary Demographic Institute, The Hague, The Netherlands
e-mail: Willekens@nidi.nl

J. Stillwell, M. Clarke (eds.), *Population Dynamics and Projection Methods*, Understanding Population Trends and Processes 4, DOI 10.1007/978-90-481-8930-4_2,

parameters from incomplete data. It identifies two steps. The first step predicts the model outcome using preliminary estimates of the parameters of the model. It is the expectation (E) step. In the second step, the maximization step (M), improved parameter estimates are obtained. The same two steps are distinguished by Rees and Wilson.

The structure of the chapter is as follows. The next section reviews basic features of multistate demographic accounts and distinguishes between accounts on the basis of the measurement of flows of different types. This is followed by a presentation of the algorithm proposed by Phil Rees for estimating the demographic parameters from incomplete data. The next section compares the algorithm proposed by Rees and the EM algorithm. A section of conclusions completes the chapter.

Multistate Demographic Accounts

Demographic accounts are tables that combine data on population stocks and flows. A population consists of people with various attributes such as age, sex, region of residence, marital status, employment status and health status. Attributes change over time and the change can be positioned in one or several time scales. Common time scales include individual time (age) and calendar time. A time scale measures the time elapsed since a reference event such as a census. Age is the duration of life or time elapsed since birth. Calendar time (e.g. t) is the time elapsed since the start of our calendar. Data on attributes and particularly data on attribute changes are generally incomplete. For some people, the attributes may not be known and, even if the attributes are known, changes may remain unnoticed or may be recorded with some delay. For example, whilst we may know thc number of people who migrate from one region to another in a time period, it is very likely that we will not know the numbers who migrate and die during the period or those who are born and migrate as infants.

The rules or principles of spatial demographic accounting state how it is possible to organize the available information in an account and how to infer missing attributes and missing information on attribute changes. The missing information may be inferred from the available data using either simple rules or complicated statistical theory. It is not uncommon that rules that have been in existence for quite some time are shown to produce results that are fully consistent with the theories of statistical inference.

The principles of accounting presented in this section are based on Rees and Wilson (1977). First, the population system needs to be defined and the boundaries of the account need to be determined. The population system for which the account is developed is generally not the world population since the origin of humankind, but a well-defined population situated in time and space. The population is divided into subpopulations on the basis of attributes of members of the population. A person with a given attribute is said to occupy a given state. The population can therefore be referred to as a multistate population. A change in attribute implies a move to a different state. Thus a resident of a given region who at time t is aged x and not employed can get a job in another region and migrate during the period between t and t+1. As a result, both the employment status and the region of residence change during the period of 1 year. Note that a person's change of state may remain unregistered for some time or even forever. A distinction is made between the occurrence of an event, which depends on the definition of the event, and the measurement of the occurrence. For instance, if a person relocates to a new address, an event occurs. However, the geographical relocation will not show up in a statistical database unless it is reported or registered in another way, and the information is used in the compilation of statistics.

A multistate population is embedded in a larger population and in history. If no exchange with the larger population is possible, the population is a closed population. In general, the population system for which a population account is prepared is an open system, unless the account refers to the whole of the world. If the population account is for one country/nation, persons may enter by birth or immigration and leave the population by death or emigration. If the population covers a particular age range, a person enters by reaching the lowest age of the age range or immigration and leaves by reaching the highest age or emigration. The account may cover a subpopulation only, e.g. students, health workers, retirees or the residents of a city or region. Moves across boundaries of an open population are sometimes termed external events as opposed to internal events that do not involve the crossing of the boundary of the population system. External events comprise exits and entries. Exits may be subdivided by destination, entries by origin.

Persons are located in space and time. Measurement methods differ substantially in the way they register the location (in space) of a person and the date (location in time) of an event. The location of a person may be measured precisely and displayed in a Cartesian coordinate system or approximately by using a grid system of regions. In a grid system, an area is divided in squares of, say, one kilometre by one kilometre. In a multiregional system, an area is divided in regions that may differ in shape and size. In multiregional population accounts, the location is given in terms of the region of residence. The date of a change in attribute or event may be measured precisely and reported in day, month and year, or approximately and reported in month or year of occurrence. Many demographic surveys report dates of events in terms of month and/or year. Relocations and other events are measured directly or indirectly. The indirect measurement is obtained by comparing regions of residence and other personal attributes at two points in time.

Direct and indirect measurements result in different data types. Data resulting from the direct measurement of events during a given interval have been referred to as movement data (*moves*). Data resulting from the indirect measurement of events have been referred to as transition data (*migrants*); they are essentially a cross-tabulation of final states (state occupied at the end of an interval) and initial states (state occupied at the beginning of an observation interval). Different data types require different methods for estimating the parameters of demographic models (see Ledent, 1980, p. 558).[1] They also lead to different accounts: *movement accounts* and *transition accounts* (Rees & Willekens, 1986).

[1]Rogers (1975) refers to the method that uses movement data as the 'option 1' method and the method that uses transition data as the 'option 2' method.

The direct measurement of events is illustrated in population registers. A population register is designed to be a continuous surveillance system and to record events as they occur. A person experiencing an event is expected to report the event right away or within a brief period of time. The date of the event (or date of registration) is recorded. The indirect measurement is illustrated in a population census that records the place of residence at two points in time (e.g. place of residence at the census and at previous census or 1 or 5 years prior to the current census). If the places of residence differ, the person has changed residence at least once. If the places are the same, a person may have left his residence and returned during the observation interval. The census may also register retrospectively the date of the change of address or the duration of stay at the current address. Demographic accounts should be able to handle different types of data and to harmonize data of different type. In developing a typology of data, it is important to distinguish between moves (also referred to as direct transitions) and transitions.

Construction of Population Accounts

Rees and Wilson (1977) consider a mixture of movement and transition data. Births and deaths are movement data and migrations are measured indirectly by comparing the region of residence at the start of a time interval and the region of residence at the end of the interval. The discrete time interval is from t to $t+T$, with T the interval length. The population at time t is the initial population and the population at $t+T$ is the final population.

The principles of demographic accounting proposed by Rees and Wilson are now briefly outlined. Let $K_i(t)$ denote the population in region i at time t and let K_{ij} denote the number of persons in region i at time t and region j at time $t+T$ ($i = 1, 2, 3, \ldots, N$ and $j = 1, 2, 3, \ldots, N$). K_{ii} denotes the number of people in region i at the beginning of the interval (at time t) who are also in i at the end of the interval. The number of residents of region i at t who die in region j during the interval is denoted by $K_{i\delta(j)}$. If j differs from i, it represents a migration followed by a death. The number of children born in region i during the interval who survive and live in region j at the end of the interval is denoted by $K_{\beta(i)j}$. Some children born during the interval die before the end of the interval. That number is denoted by $K_{\beta(i)\delta(j)}$. Rees and Wilson (1977, pp. 22 ff.) distinguish major flows and minor flows. Major flows involve one demographic event only (in addition to survival). They are K_{ij}, $K_{\beta(i)i}$ and $K_{i\delta(i)}$. The others are minor flows.

The development of the accounts table consists of five steps:

1. Assemble the available data in the account.
2. Obtain preliminary/improved estimates of the population at risk.
3. Estimate birth and death rates for each region.
4. Obtain estimates of the minor flows.
5. Go to step 2.

The population at risk in a given region measures the total time spent in that region by all persons combined included in the account. The exposure time (population at risk) is determined for the events of birth and death, but not for migration. Since migration is measured by comparing the initial region of residence (at t) and the final region of residence (at time $t+T$), the base population is considered a suitable approximation of the population at risk (Rees & Wilson, 1977, p. 34).

The estimation of the population at risk of an event (birth or death) from the available data requires assumptions about the timing of the events in the interval from t to $t+T$. It is assumed that the K_{ij} persons who are in i at t and in j at $t+T$ and who experience a single event during the interval, experience the event in the middle of the interval, hence at time $t+0.5T$. As a consequence, a surviving migrant is exposed half a period in the region of origin i and half a period in the region of destination j. The $K_{i\delta(i)}$ persons who die in the region in which they reside at time t die in the middle of the interval. They also contribute half a period to the total exposure in region i. The $K_{i\delta(j)}$ persons who migrate to another region and subsequently die in that region contribute $0.25T$ to the total exposure in region i and $0.25T$ to the total exposure in region j. It implies the assumption that death occurs in the middle of the period and the migration halfway through the first subperiod. The $K_{\beta(i)i}$ children who are born in region i and are in the same region at the end of the interval are assumed to be born in the middle of the interval. They contribute $0.5T$ to the exposure time. The $K_{\beta(i)j}$ newborns who migrate to another region contribute $0.25T$ to the region of birth and $0.25T$ to the region of destination. The $K_{\beta(i)\delta(i)}$ newborns who die in their region of birth contribute $0.25T$ to the exposure time in the region of birth, implying the assumption that they are born in the middle of the interval and die in the middle of the second sub-period. The $K_{\beta(i)\delta(j)}$ newborns who die in a different region are assumed to contribute $0.125T$ to the total exposure in the region of birth and $0.125T$ to the exposure in the region of destination (and death). This implies the assumption that the child is born in the middle of the interval of length T, dies in middle of the second subinterval of length $0.5T$, and migrates in the middle of the first part of the second interval, which has a length of $0.125T$.[2]

The person-years of exposure in region i to the event of birth is:

$$\begin{aligned} PY_i^B &= T\,K_{ii} + 0.5T\sum_{j\neq i} K_{ij} + 0.5T\sum_{j\neq i} K_{ji} + 0.5T\,K_{i\delta(i)} \\ &\quad + 0.25T\sum_{j\neq i} K_{i\delta(j)} + 0.25T\sum_{j\neq i} K_{j\delta(i)} \end{aligned} \tag{2.1}$$

[2]In the context of event history analysis, Yamaguchi (1991) uses a similar reasoning to arrive at estimates of exposure time.

And the person-years of exposure in region i to the event of death is:

$$PY_i^D = PY_i^B + 0.5T\,K_{\beta(i)i} + 0.25T\,K_{\beta(i)\delta(i)} + 0.25T\sum_{j\neq i} K_{\beta(i)j} + 0.25T\sum_{j\neq i} K_{\beta(j)i} + 0.125T\sum_{j\neq i} K_{\beta(i)\delta(j)} + 0.125T\sum_{j\neq i} K_{\beta(j)\delta(i)} \quad (2.2)$$

Several of these quantities are not available. The equations are therefore rewritten to express the person-years in terms of observed quantities and unknown quantities. The following terms are usually available directly from the data (Rees & Wilson, 1977, p. 23): the initial population in region i, K_{i*}, the total number of births, $K_{\beta(i)*}$, the total number of deaths, $K_{*\delta(i)}$, the numbers of surviving migrants, K_{ij} $(i\neq j)$, and migrating infants, $K_{\beta(i)j}$ $(i\neq j)$. The approximate person-years expressed in these terms are:

$$PY_i^B = T\,K_{i*} - 0.5T\,K_{*\delta(i)} + 0.5T\sum_{j\neq i}(K_{ji} - K_{ij}) \quad (2.3)$$

for births and for deaths it is

$$PY_i^D = PY_i^B + 0.5T\,K_{\beta(i)*} \quad (2.4)$$

The birth rate in region i is obtained by dividing the number of births by the person-years of exposure:

$$b_i = K_{\beta(i)*}/PY_i^B \quad (2.5)$$

The rate is an occurrence-exposure rate because it is obtained by dividing the occurrences (number of births) by the exposure time. Since exposure time is measured in person-years, the rate reflects the intensity of the event in a year. In most applications the time unit is a year, but in some applications of multistate models it is a month or even a day. The death rate is obtained in similar way as:

$$d_i = K_{*\delta(i)}/PY_i^D \quad (2.6)$$

To illustrate the development of a population account, I use the example provided by Rees and Wilson. Three regions are considered: the West Riding of Yorkshire in the UK (region 1), the rest of England and Wales (region 2), and the rest of the world (region 3). The initial step in developing the population account is to fill the accounts table with available data. Table 2.1 shows the known population stocks and flows.

The approximate person-years based on the available data and the associated birth and death rates are estimated using the equations shown above. For the West Riding, the average annual birth rate is 18.9 per thousand and the death rate 12.3 per thousand. For the rest of England and Wales, the annual birth rate is 19 per

Table 2.1 Population account table for the West Riding of Yorkshire, 1961–1966, known flows

		Population 1966			Deaths 1961–1966			
		1	2	3	1	2	3	Total
Population 1961	1	–	168,207	45,089	–	–	–	3,650,586
	2	137,183	–	841,621	–	–	–	42,453,962
	3	58,804	1,026,175	–	–	–	–	–
Births in 1961–1966	1	–	7,492	3,808	–	–	–	333,135
	2	7,003	–	57,890	–	–	–	3,924,421
	3	2,617	44,250	–	–	–	–	–
Total		–	–	–	226,694	2,537,636	–	–

Source: Rees and Wilson (1977, p. 57).

thousand and the death rate is 11.7 per thousand. The death rates are employed to calculate the following minor flows: the $K_{i\delta(j)}$ migrants who die before the end of the interval, the $K_{\beta(i)\delta(j)}$ children born in region i who migrate to region j and die before the end of the interval, and the $K_{\beta(i)\delta(i)}$ children born in region i who die in that region before the end of the interval. To determine how many migrants from i to j die in j before the end of the interval, the number of migrants, K_{ij}, is multiplied by a function of the rate of dying during a period of $0.25T$:

$$K_{i\delta(j)} = \frac{0.5T\ {}_1d_j}{1 - 0.25T\ {}_1d_j} K_{ij} \tag{2.7}$$

where ${}_1d_j$ is the annual death rate in region j. The number of children born in i who die in j is:

$$K_{\beta(i)\delta(j)} = \frac{0.25T\ {}_1d_j}{1 - 0.125T\ {}_1d_j} K_{\beta(i)j} \tag{2.8}$$

The number of children born in i who die in i is:

$$K_{\beta(i)\delta(i)} = \frac{0.25T\ {}_1d_i}{1 - 0.125T\ {}_1d_i} K_{\beta(i)i} \tag{2.9}$$

where:

$$K_{\beta(i)i} = K_{\beta(i)*} - \sum_{j \neq i} K_{\beta(i)j} - K_{\beta(i)\delta(i)} - \sum_{j \neq i} K_{\beta(i)\delta(j)} \tag{2.10}$$

Rearrangement gives:

$$K_{\beta(i)\delta(i)} = \frac{0.25T\ {}_1d_i}{1 + 0.125T\ {}_1d_i} \left[K_{\beta(i)*} - \sum_{j \neq i} K_{\beta(i)j} - \sum_{j \neq i} K_{\beta(i)\delta(j)} \right] \tag{2.11}$$

Once these minor flows are estimated, better estimates can be obtained of the person-years at risk and the birth and death rates. Iteration results in final estimates of the missing flows in the population account. Table 2.2 shows the final account.

Demographic Accounting Method and EM Algorithm Compared

The final account is obtained by an iterative procedure involving three basic steps. The first is to obtain the population at risk (person-years of exposure) using a demographic accounting equation. In the second step, the births and death rates are obtained by dividing the numbers of events by the populations at risk. In the third step, these updated rates are entered into a model derived from the demographic accounting equation that predicts the missing data. The rates that are produced in the

Table 2.2 Population account table for the West Riding of Yorkshire, 1961–1966, final estimates

		Population 1966			Deaths 1961–1966			
		1	2	3	1	2	3	Total
Population 1961	1	3,220,324	168,207	45,089	210,558	5,004	1,404	3,650,586
	2	137,183	39,056,927	841,621	4,271	2,388,922	25,038	42,453,962
	3	58,804	1,026,175		1,831	30,528		–
Births in 1961–1966	1	311,781	7,492	3,808	9,885	111	59	333,135
	2	7,003	3,746,147	57,890	108	112,418	855	3,924,421
	3	2,617	44,250		40	653		–
Total		3,737,712	44,049,198		226,694	25,37,636	–	–

Source: Rees and Wilson (1977, p. 72).

second step are occurrence-exposure rates. An occurrence-exposure rate is a ratio of the number of events experienced by a group of people during a given period and the exposure time during the same period.

The EM algorithm is also based on a model of the data and is an iterative procedure too. The algorithm was developed by Dempster, Laird, and Rubin (1977) and is currently the main method for maximum likelihood estimation in the presence of missing data. The EM algorithm uses a probability model to describe the event of interest. The type of model depends on the type of event. Births, deaths and migrations (moves) are described by Poisson models or by an extension of the Poisson model such as the negative binomial model. The selection of a model determines the probability distribution of the expected number of events. If the Poisson model is used, it is assumed that the number of events during a time interval follows a Poisson distribution. Transitions in discrete time (e.g. migrants, see before) are described by multinomial probability models. The models listed here are standard models in probability theory and they are described in any introductory textbook. For a thorough review of the EM algorithm, see McLachlan and Krishnan (1997). For a discussion of the EM algorithm in the context of migration analysis, see Willekens (1999). The demographic accounting method is not based explicitly on a probability model. The model is implicit, however, in the estimation of the person-years of exposure and, more particularly, in the assumption that events occur in the middle of the risk period. That assumption is consistent with the assumption that events are uniformly distributed during the risk period. The uniform distribution is one of the common probability distributions documented in probability theory. When the occurrence of events follows a uniform distribution, then the expected waiting time to an event, provided it occurs during a given time interval, is half the interval. Traditionally the uniform distribution is (implicitly) assumed in demographic models of event sequences. The Poisson distribution, which implies an exponential distribution of events during the risk period, is (implicitly) assumed in most statistical models of event sequences. The distribution assumes that the rate at which events occur is constant.

In the EM algorithm, the parameters of the distribution are estimated from the data using the maximum-likelihood method. The method maximizes the likelihood of the data given the model. Maximum-likelihood estimates have a number of statistical properties. In the demographic accounting method, the statistical properties of the parameter estimation method are not studied. The properties can be inferred, however. Andersen and Keiding (2002), Andersen and Perme (2008) and others show that the occurrence-exposure rate estimated by dividing an event count by an exposure time agrees well with the Nelson-Aalen estimator *provided the rate at which events occur is constant*. The Nelson-Aalen estimator is a non-parametric estimator of the cumulative rate of the transition/hazard rate. The estimator is part of the toolkit of statistical survival analysis and event history analysis and is well-documented in the literature. Note that the occurrence-exposure rate estimated by the demographic accounting method is not consistent with the Nelson-Aalen estimator because of the assumption of uniform distribution of events.

The EM algorithm is a generic method that offers maximum likelihood solutions when data are incomplete. It reformulates the incomplete-data problem as a complete-data problem and estimates the parameters of the probability model in two steps. The first step of the algorithm infers the missing data using preliminary estimates of the parameters of the model. It is the expectation (E) step. In the second step, the 'complete' data and the maximum likelihood method are used to improve on the parameter estimates. It is the maximization (M) step. These two steps, that are the main characteristic of the algorithm, are also present in the Rees-Wilson account-based method. The first step predicts event occurrences (event counts) on the basis of available data and preliminary estimates of exposure time. The second step estimates event rates (fertility, mortality and migration rates) by dividing event occurrences by exposure time. The third step repeats the first step to produce improved predictions of event counts. The similarity between these two characteristic steps is the reason why the Rees-Wilson method is referred to as 'EM algorithm *avant-la-lettre*'.

Conclusion

The demographic accounting method, pioneered by Phil Rees, is an EM algorithm *avant-la-lettre*. It distinguishes the two basic steps in model development with incomplete data: the prediction of the model outcome from preliminary parameter estimates and the improved estimation of the parameters from the 'complete' data. It results in parameter estimates that are optimal in some way given the data. The parameters have interesting statistical properties that became known only much later after the development of the statistical theory of counting processes.

Phil Rees showed convincingly that a demographic account represents an appropriate basis for demographic modelling (see e.g. Rees, 1979). Since accounts must balance, the accounting framework assures consistency between the flows and the stocks from which the model parameters are derived. The power of account-based models has often been demonstrated, recently in a major report on ethnic population projections for the UK (Rees, Norman, Wohland, & Boden, 2010). The observation that account-based models and models that are based on statistical inference share important features may lead to renewed research that further strengthen the position of the accounting method in modelling.

References

Andersen, P. K., & Keiding, N. (2002). Multi-state models for event history analysis. *Statistical Methods in Medical Research, 11*, 91–115.

Andersen, P. K., & Perme, M. P. (2008). Inference for outcome probabilities in multi-state models. *Lifetime Data Analysis, 14*, 405–431.

Dempster, A. P., Laird, M. M., & Rubin, D. B. (1977). Maximum likelihood fromincomplete data via the EM algorithm (with discussion). *Journal of the Royal Statistical Society B, 39*, 1–38.

Ledent, J. (1980). Multistate life tables: Movement versus transition perspectives. *Environment and Planning A, 12*(5), 533–562.
McLachlan, G. J., & Krishnan, T. (1997). *The EM algorithm and extensions*. New York: Wiley.
Rees, P. H. (1979). Regional population projection models and accounting methods. *Journal of the Royal Statistical Society A, 142*, 223–255.
Rees, P., Norman, P., Wohland, P., & Boden, P. (2010). *Report on ethnic population projections for the UK*. School of Geography, University of Leeds, Leeds.
Rees, P. H., & Willekens, F. (1986). Data and accounts. In A. Rogers & F. Willekens (Eds.), *Migration and settlement: A multiregional comparative study* (pp. 19–58). Reidel Press: Dordrecht.
Rees, P., & Wilson, A. (1977). *Spatial population analysis*. London: Edward Arnold.
Rogers, A. (1975). *Introduction to multiregional mathematical demography*. New York: Wiley.
Willekens, F. J. (1999). Modeling approaches to the indirect estimation of migration flows: From entropy to EM. *Mathematical Population Studies, 7*(3), 239–278.
Yamaguchi, K. (1991). *Event history analysis*. Newbury Park, CA: Sage Publications.

Chapter 3
MULTIPOLES: A Revised Multiregional Model for Improved Capture of International Migration

Marek Kupiszewski and Dorota Kupiszewska

Introduction

It has been argued on many occasions (e.g. in Kupiszewski & Kupiszewska, 1997, 2008; Kupiszewski, 2002) that there is a need for models of multiregional population dynamics capable of simultaneously producing consistent results at various spatial levels: regional, national and supranational. Kupiszewski and Kupiszewska (2008) outlined the changing needs for modelling complex population systems, the impact of migration data availability and quality on modelling strategies and methodological developments making the construction of such complex models feasible. The authors pointed out that in the late 1980s and early 1990s a need emerged to model large, multinational population systems at the sub-national level. This need continues today for a number of reasons. First, the development of the European Union (EU) regional policy requires the provision of regional indicators and forecasts (including migration and population) as an input to policy making. Second, some EU funding allocations are dependent on estimates of the stock of immigrants. Third, the globalization of migration flows has made international migration relevant for sub-national regional labour markets. Simultaneously, in many European countries, international migration has become an important, or even the dominant, factor shaping population dynamics. This is illustrated in Table 3.1 using example data for the UK and Poland for 1991 and 2008.

In the late 1980s, two main approaches to supra-national population forecasts could be distinguished: (i) forecasts prepared for a set of countries, but individually for each country, with international migration treated as an external factor; (ii) simultaneous modelling of the populations of a system of countries, but without the regional breakdown. The first response to the changing needs was by Rees and colleagues (Rees, Stillwell, & Convey, 1992; Rees, 1996a). The *E*uropean *C*ommunity *POP*ulation projection model (*ECPOP*) was designed to project the populations of the 12 member states of the European Community for the NUTS

M. Kupiszewski (✉)
Institute of Geography and Spatial Organization, Polish Academy of Sciences, Twarda 51/55, 00-818 Warszawa, Poland
e-mail: m.kupisz@twarda.pan.pl

J. Stillwell, M. Clarke (eds.), *Population Dynamics and Projection Methods*, Understanding Population Trends and Processes 4, DOI 10.1007/978-90-481-8930-4_3,

Table 3.1 Illustration of the changing role of international migration in population dynamics

	Poland		United Kingdom	
	1991	2008	1991	2008
Natural increase ('000s)	142.0	35.1	146.1	214.7
Net international migration incl. corrections ('000s)	−15.9	−14.9	27.3	240.7

Source: Eurostat.

1 regions. In addition to fertility and mortality scenarios, it allowed scenarios to be set for three classes of migration: interregional (intra-member state) migration, international intra-Community migration and extra-Community migration. In many ways, Rees' model was a major improvement in the practice of population projection. From the methodological point of view, it was a state of the art model based on population projection theory which allowed for a coherent and unified treatment of supra-national but regionally disaggregated populations, based on Rogers' (1975) concepts that had been developed earlier for multiregional models.

Rees' model gave us the idea for the development of the *MULTIPOLES* model, initially to be used for projecting populations of Central and Eastern European countries. However, there are numerous differences between *ECPOP* and *MULTIPOLES,* perhaps the most important being that unlike *ECPOP*, the *MULTIPOLES* model is based on movement type population accounts instead of the transition accounts. The former model was written with the assumption that most of the migration data would be coming from censuses. In Central Europe, population registers are the main source of migration data, so the use of movement population accounts seemed more suitable.

Since its inception, the *MULTIPOLES* model has been revised and improved several times. It was used in a number of research projects (Kupiszewski & Kupiszewska, 1997, 1999; Kupiszewski, 2001, 2002; Bijak, Kupiszewska, Kupiszewski, & Saczuk, 2005; Bijak, Kupiszewska, Kupiszewski, Saczuk, & Kicinger, 2007; Bijak, Kupiszewska, & Kupiszewski, 2008; Bijak & Kupiszewski, 2008) and is currently undergoing a comprehensive re-write to meet the requirements of the DEMIFER project.[1] In the next section, we present the first formulation of the model, as used in the application presented later (it still exists in the current version of *MULTIPOLES*, next to the new options which have been added to increase the flexibility of input data specification and scenario formulation).

MULTIPOLES: A *MULTI*state *PO*pulation Model for Multi*LE*vel *S*ystems

The *MULTIPOLES* is a cohort-component, hierarchical, multiregional, supranational model of population dynamics. It may be used for forecasts, projections and

[1] DEMIFER stands for Demographic and Migratory Flows Affecting European Regions and Cities and is funded by the European Union via the ESPON programme.

simulations. The population is disaggregated into sexes and 18 five-year age groups, i.e. 19 projection cohorts with the youngest cohort being the infant cohort (children born during the projection interval) and the cohort aged 85+ being the oldest one. Geographically, the population is disaggregated into countries and regions. The rates appearing in the accounts are defined as the number of events (deaths, migration or births) in a projection period divided by the population at risk, calculated as an arithmetic average of the population of the projection cohort at the beginning and at the end of the projection period. Migration is handled on three levels, as in the *ECPOP* model:

- inter-regional intra-national migration within each country;
- inter-regional international migration within the system;
- net migration from the 'Rest of the World' to each country within the system.

The structure of the *MULTIPOLES* model is presented in Fig. 3.1.

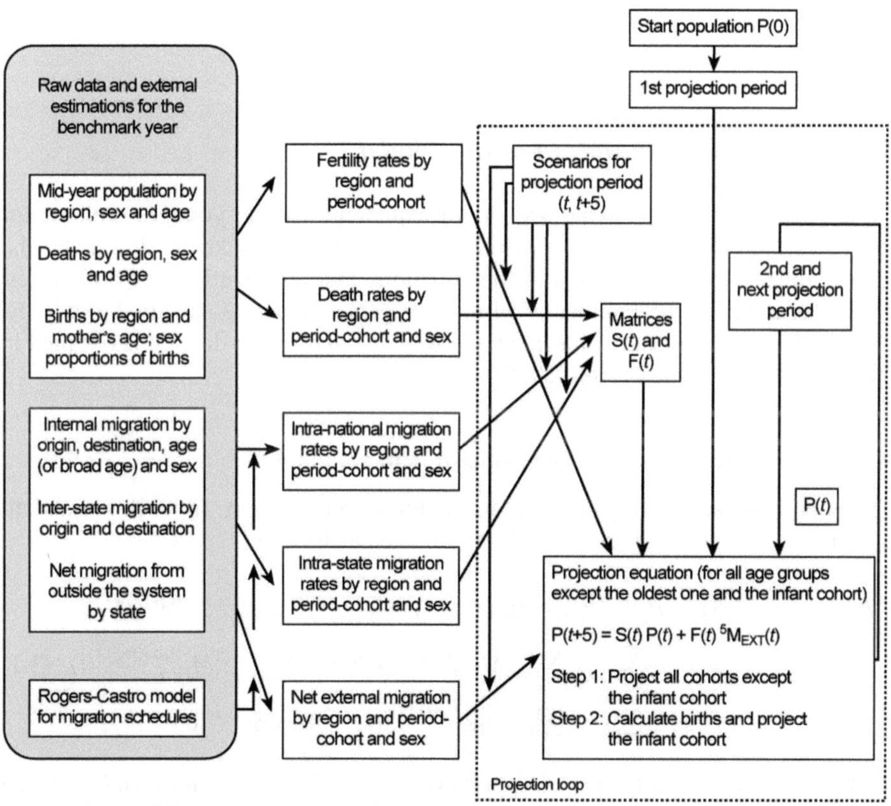

Fig. 3.1 The structure of the *MULTIPOLES* model

Definition of the Population System and Notation

Each region is identified by a pair of indexes (*is,ir*), where *is* (state *i*) denotes the number of a country and *ir* (region *i*) the number of a region in this country. Such notation guarantees an elasticity of the model and the possibility to redefine both the number of countries modelled as well as the number of regions in each country easily. When we refer to events occurring in age group *a* over period (*t,t*+5), we have in mind the period-cohort measurement plan; that is the events concerning persons at the age (*a,a*+5) at time *t* which took place over period (*t,t*+5). In the formulae, the following notation has been used:

t – time
g – sex (f – female, m – male)
a – age group (covering persons at the age from *a* to *a*+5 years)
00 – index of the youngest cohort (children born during the projection interval)
A+ – index of the oldest, open-ended age group, covering persons of age *A* or more (85+)
ir, jr – regions
is, js – countries.

Mathematical Formulation of the Model

In this section we present the accounting equations for the youngest cohort, for the oldest cohort and for the remaining cohorts. For the sake of simplicity, the gender index has been omitted in the equations that are identical for both sexes. In the formula for the number of births, the gender index is specified explicitly. Only the key equations are presented; the way they have been derived is outlined in (Kupiszewska & Kupiszewski, 2005).

Accounting Equations for Age Groups $a = 0, 5, \ldots, A\text{–}5$

The population accounting equation for region *ir* in country *is*, for each sex and all age groups except 00 and *A*+, may be formulated as follows:

$$\begin{aligned} P_{a+5}{}^{(is,ir)}(t+5) = {} & P_a{}^{(is,ir)}(t) - {}^5D_a{}^{(is,ir)}(t) - \textstyle\sum_{jr \neq ir} {}^5M_{\mathrm{IR}a}{}^{(is,ir)(is,jr)}(t) \\ & - \textstyle\sum_{js \neq is} \sum_{jr} {}^5M_{\mathrm{IS}a}{}^{(is,ir)(js,jr)}(t) + \textstyle\sum_{jr \neq ir} {}^5M_{\mathrm{IR}a}^{(is,jr)(is,ir)}(t) \quad (3.1) \\ & + \textstyle\sum_{js \neq is} \sum_{jr} {}^5M_{\mathrm{IS}a}^{(js,jr)(is,ir)}(t) + {}^5M_{\mathrm{EXT}a}{}^{(is,ir)}(t) \end{aligned}$$

To distinguish clearly two categories of inter-regional migration (internal and inter-country), separate variables have been used (M_{IR} and M_{IS} respectively). Thus, the variables appearing in the above equation have the following meaning:

$P_a{}^{(is,ir)}(t)$	– Population in age group a in region ir in country is at time t, that means at the beginning of a projection step
$P_a{}^{(is,ir)}(t+5)$	– Population in age group a in region ir in country is at time t+5, that means at the end of a projection step
${}^5D_a{}^{(is,ir)}(t)$	– Deaths in age group a in region ir in country is over period (t,t+5)
${}^5M_{\mathrm{IR}a}{}^{(is,ir)(is,jr)}(t)$	– Interregional migration from region ir to region jr in country is in age group a over period (t,t+5)
${}^5M_{\mathrm{IS}a}{}^{(is,ir)(js,jr)}(t)$	– International migration from region ir in country is to region jr in country js in age group a over period (t,t+5)
${}^5M_{\mathrm{EXT}a}{}^{(is,ir)}(t)$	– Net migration from the Rest of the World in region ir in country is in age group a over period (t,t+5)

The number of deaths and inter-regional migration events may be expressed as a product of population at risk and, respectively, death rates $d_a^{(is,ir)}(t)$, internal out-migration rates $m_{\mathrm{IR}a}^{(is,ir)(is,jr)}(t)$ and inter-regional international emigration rates $m_{\mathrm{IS}a}^{(is,ir)(js,jr)}(t)$.

After transformations and using the matrix notation, the equation for calculating population at the end of a projection period may be formulated as follows:

$$\mathbf{P}_{a+5}(t+5) = [\mathbf{I}+0.5\mathbf{M}_a(t)]^{-1}\cdot[\mathbf{I}-0.5\mathbf{M}_a(t)]\cdot\mathbf{P}_a(t)+[\mathbf{I}+0.5\mathbf{M}_a(t)]^{-1}\cdot{}^5\mathbf{M}_{\mathrm{EXT}a}(t) \tag{3.2}$$

where $\mathbf{P}_a(t)$ is a vector of regional stocks of population at age group a at time t in all the countries, $\mathbf{M}_a(t)$ is a matrix dependent on death rates, internal out-migration rates and on international emigration rates (for the definition of $\mathbf{M}_a(t)$, see Kupiszewska & Kupiszewski, 2005). Vector ${}^5\mathbf{M}_{\mathrm{EXT}a}(t)$ contains net migration from the Rest of the World to all individual regions of the modelled system in the period (t,t+5), expressed as absolute numbers rather than the rates. **I** is the identity matrix.

Accounting Equations for the Youngest Cohort

Age group 00, the youngest cohort, comprises children born during the current step of the projection. The accounting equation for this age group has the form:

$$\begin{aligned} P_0{}^{(is,ir)}(t+5) = {}& {}^5B^{(is,ir)}(t) - {}^5D_{00}{}^{(is,ir)}(t) - \textstyle\sum_{jr\neq ir}{}^5M_{\mathrm{IR}00}{}^{(is,ir)(is,jr)}(t) + \\ & - \textstyle\sum_{js\neq is}\sum_{jr}{}^5M_{\mathrm{IS}00}{}^{(is,ir)(js,jr)}(t) + \sum_{jr\neq ir}{}^5M_{\mathrm{IR}00}{}^{(is,jr)(is,ir)}(t) + \\ & + \textstyle\sum_{js\neq is}\sum_{jr}{}^5M_{\mathrm{IS}00}{}^{(js,jr)(is,ir)}(t) + {}^5M_{\mathrm{EXT}00}{}^{(is,ir)}(t) \end{aligned} \tag{3.3}$$

where ${}^5B^{(is,ir)}(t)$ denotes births in region ir in country is over period (t,t+5).

Using the matrix notation and transforming the equation we arrive at the following equation:

$$\mathbf{P}_0(t+5) = [\mathbf{I}+0.5\mathbf{M}_{00}(t)]^{-1}[{}^5\mathbf{B}(t) + {}^5\mathbf{M}_{\mathrm{EXT}00}(t)] \tag{3.4}$$

where ${}^{5}\mathbf{B}(t)$ is a vector of births of a given sex over the period $(t,t+5)$, i.e. a vector which elements are ${}^{5}B^{(is,ir)}(t)$. The remaining variables have a similar meaning as in Equation (3.2), but are calculated for the youngest cohort.

The number of births of sex g, ${}^{5}B_{g}{}^{(is,ir)}(t)$, can be expressed as:

$$ {}^{5}B_{g}^{(is,ir)}(t) = 0.5 f_{g}^{is} \sum_{a} {}^{5}b_{a}^{(is,ir)}(t)[P_{\mathrm{af}}^{(is,ir)}(t) + P_{(a+5)\mathrm{f}}^{(is,ir)}(t+5)] \tag{3.5} $$

where ${}^{5}b_{a}^{(is,ir)}(t)$ is fertility rate of females in age group a in region ir in country is calculated over period $(t,t+5)$, f_{g}^{is} is the proportion of newborn children in country is who are of sex g and the sum goes through all fertile age goups.

Accounting Equations for the Oldest Cohort (A+)

Accounting equations for the oldest age group can be obtained taking into account that:

$$ \mathbf{P}_{A+}(t+5) = \mathbf{P}_{(A+5)+}(t+5) + \mathbf{P}_{A}(t+5) \tag{3.6} $$

Proceeding as before, we obtain:

$$ \begin{aligned} \mathbf{P}_{A+}(t+5) = {} & [\mathbf{I} + 0.5\mathbf{M}_{A+}(t)]^{-1} \cdot [\mathbf{I} - 0.5\mathbf{M}_{A+}(t)] \cdot \mathbf{P}_{A+}(t) \\ & + [\mathbf{I} + 0.5\mathbf{M}_{A+}(t)]^{-1} \cdot {}^{5}\mathbf{M}_{\mathrm{EXT}A+}(t) + \\ & + [\mathbf{I} + 0.5\mathbf{M}_{A-5}(t)]^{-1} \cdot [\mathbf{I} - 0.5\mathbf{M}_{A-5}(t)] \cdot \mathbf{P}_{A-5}(t) \\ & + [\mathbf{I} + 0.5\mathbf{M}_{A-5}(t)]^{-1} \cdot {}^{5}\mathbf{M}_{\mathrm{EXT}(A-5)}(t) \end{aligned} \tag{3.7} $$

Projection Equations

Putting together the equations presented in the three previous sections, we arrive at the following set of projection equations covering all the age groups:

$$ \mathbf{P}_{0}(t+5) = \mathbf{F}_{00}(t)[{}^{5}\mathbf{B}(t) + {}^{5}\mathbf{M}_{\mathrm{EXT}00}(t)] \tag{3.8} $$

$$ \mathbf{P}_{a+5}(t+5) = \mathbf{S}_{a}(t)\mathbf{P}_{a}(t) + \mathbf{F}_{a}(t){}^{5}\mathbf{M}_{\mathrm{EXT}a}(t) \tag{3.9} $$

$$ \begin{aligned} \mathbf{P}_{A+}(t+5) = {} & \mathbf{S}_{A+}(t)\mathbf{P}_{A+}(t) + \mathbf{F}_{A+}(t){}^{5}\mathbf{M}_{\mathrm{EXT}A+}(t) + \mathbf{S}_{A-5}(t)\mathbf{P}_{A-5}(t) \\ & + \mathbf{F}_{A-5}(t){}^{5}\mathbf{M}_{\mathrm{EXT}(A-5)}(t), \end{aligned} \tag{3.10} $$

$$ \text{where } \mathbf{S}_{a}(t) = [\mathbf{I} + 0.5\mathbf{M}_{a}(t)]^{-1}[\mathbf{I} - 0.5\mathbf{M}_{a}(t)] \text{ and } \mathbf{F}_{a}(t) = [\mathbf{I} + 0.5\mathbf{M}_{a}(t)]^{-1} \tag{3.11} $$

The matrix $\mathbf{S}$ may be interpreted as a matrix of survival coefficients for the population present in the system at time t, whereas $\mathbf{F}$ as a matrix of survival coefficients for the population which joined the system through births or immigration from the Rest of the World in the projection period $(t,t+5)$. The equations look like the traditional multiregional population projection equations for a single country, the difference is in how matrix $\mathbf{M}_a$ is constructed, i.e. that two types of inter-regional migration flows are involved: $\mathbf{M}_a(t)$ depends not only on internal out-migration rates but also on the rates of emigration to the regions in the other countries. Moreover, vector $\mathbf{M}_{\mathrm{EXT}}$ contains not all international migration but only the international migration to and from the countries not belonging to the modelled system.

An Application of *MULTIPOLES* and an Assessment of the Error

There are various ways to assess model performance, one of which is to measure the *ex-post* forecasting error and compare it with the errors of selected other forecasts. The *ex-post* error assessment is possible only when the time elapsed from the base year of the forecast to the year for which the error calculation takes place is long enough, so that real data become available. For this reason we chose as a test case one of the first applications of *MULTIPOLES* – namely the forecast of the elderly population in Central and Eastern Europe (Kupiszewski & Kupiszewska, 1999).

Geographical and Temporal Scope of the Projection

Geographically, the forecast covered 14 countries of Central and Eastern Europe (Austria, Belarus, Czech Republic, Estonia, Germany, Hungary, Latvia, Lithuania, Moldova, Poland, Romania, Slovenia, Slovakia and Ukraine) with 154 regions altogether. The countries were selected based on two criteria: geographic location and migration interactions. Germany and Austria were therefore included, despite apparent differences in the level of economic and social development between these two countries and the former communist block countries. The forecast was based on data for 1994 which, roughly speaking, represented the situation in the Central and East European countries in the transition period. The forecast covered 25 years (1994–2019) and included simple scenarios relating to changes of life expectancy at birth, total fertility rates, migration rates for internal and international intra-system migration, and net migration numbers for international migration with the Rest of the World.

The administrative division of the states had an impact on the way the data were collected and consequently on the way the projection was conducted. Whenever it was feasible, the first level of administrative division was used. The number of regions in the countries ranged from one to 49. Some countries, namely all the Baltic states and Slovenia, have not been subdivided any further due to their small populations. For Belarus, Moldova and Ukraine, regional data were not available.

Data Collection – Problems and Solutions

Population data concerning well-defined events such as births and deaths and somewhat less well-defined migration are routinely collected by national statistical offices. In practice, we have good quality data on births and deaths and much worse data on migration and on population stocks. Data on migration, both internal and international, are often underestimated (e.g. due to avoidance of reporting).

The *MULTIPOLES* model required the following data: mid-year population stocks and, for the base year, the number of deaths by age (18 five-year age groups), sex and region; births by sex of the child born, region and age of the mother (from 15–19 years to 45–49 years; births from mothers younger than 15 and older than 49 years were counted in the adjacent age groups); a full migration matrix (origin-destination-age-sex, ODAS) for internal migration; an origin-destination matrix of international migration between the modelled countries; and net migration for exchanges between each of the countries within the system and the Rest of the World.

Population Stocks, Births and Deaths

Data on mortality and fertility are usually available and of good quality. The availability of stock data is good, however in some cases it was necessary to estimate the regional age distribution for the oldest age groups, for which the national age distributions were used as benchmarks. The quality of data on stocks of population is usually directly affected by the poor quality of migration registration. This is clearly demonstrated by the magnitudes of post-census corrections, which in the 2000 Census round were, for example, –395,600 persons for Poland and –558,200 persons for Romania (NIDI, 2004).

Migration

Data on both internal (sub-national) and international migration were more difficult to obtain. Ideally, origin-destination-age-sex data were sought for internal and international intra-system migration, and net international migration by age, sex and country for international migration from outside the system (the Rest of the World).

Internal Migration

The full (ODAS) sets of data on internal migration were available for only three countries: Poland, Romania and the Czech Republic. For the other countries it was necessary to reconstruct the full matrix of flows. The reconstruction was based on the concept of the migration cube – a three dimensional array of migration flow data. The dimensions of the cube are origin, destination and age. The array is estimated for each sex separately. In many cases only faces of this cube were available: the matrices representing migration by origin and destination, by origin and age, and by destination and age. Willekens, Por, and Raquillet (1981) elaborated algorithms allowing for the estimation of the entire cube from the marginal values. Such algorithms were used for the estimations when necessary. The process of the reconstruction of the data should not have introduced any significant errors. A more important problem was the comparability of data on internal migration (Poulain, 1994; Rees & Kupiszewski, 1999), which was far from good. In this study the data were used 'as is' and no attempt was made to bring them to a common denominator, which is a major task in itself.

International Migration

The statistics on international migration are the main source of uncertainty. There is ample literature describing the problems with the European data on international migration. The long history of efforts to solve them has been summarised by Kelly (1987) and Herm (2006). Much work on the comparability and usability of international migration data, mainly in the EU, has been done by Poulain and his colleagues (Poulain, Debuisson, & Eggerickx, 1991; Poulain, 1993, 1996) and recently within the *THESIM*[2] project (Nowok, Kupiszewska, & Poulain, 2006; Kupiszewska & Nowok, 2008) as well as the *PROMINSTAT*[3] project (Kupiszewska, Kupiszewski, Marti, & Ródenas, 2010).

The preparation of international migration data for the *MULTIPOLES* model may be split into two tasks: (i) estimation of the migration between the states covered in the study and (ii) estimation of the net migration from the outside of the system to each of the countries inside the system.

Intra-system International Flows

If we neglected the issue of the quality and comparability of data, it would be most appropriate from the methodological point of view to use data on emigration by age and destination; that is data provided by the sending countries. Using sending countries' data allows for the consistent calculation of emigration rates. However, the data on international migration collected by the sending and the receiving countries differ enormously (Kupiszewski, 1996; Kupiszewska & Nowok, 2008). That means that we may have only a very vague idea of the magnitude of international migration. The reasons for such discrepancies, such as differences in definitions, registration avoidance and different administrative procedures, have been discussed in a number of publications, for example Kupiszewska and Nowok (2006) and Kupiszewska et al. (2010), and will not be reiterated in detail here.

The problem was solved by taking the maximum of the values reported by sending and receiving countries. We chose this option, mainly due to our belief that international migration is underestimated in official statistics and therefore the larger of the two numbers is likely to estimate the migration flow size more accurately. At the time of preparing the data for the forecast we did not have enough

[2] *THESIM* stands for *T*owards *H*armonised *E*uropean *S*tatistics on *I*nternational *M*igration.

[3] *PROMINSTAT* stands for *PRO*moting Comparative Quantitative Research in the Field of *M*igration and *IN*tegration. Both projects were conducted within the EU Framework 6 Program, in 2004–2005 (*THESIM*) and 2007–2010 (*PROMINSTAT*).

information about international migration statistics in individual countries to devise a more country-specific approach. Another simple option would be to rely mainly on the data from the receiving countries, as practiced in some other studies (Raymer, 2008). The argument for using the latter solution is that, in general, immigration data are of better quality than emigration data. In our study, the under-coverage of immigration in some receiving countries (e.g. Poland, Slovakia) was huge, so using these data was not appropriate. Had we undertaken our study today, we would probably try to estimate international migration flows using the methodology developed within the *MIMOSA*[4] project. In the *MIMOSA* approach, all available flow data are taken into account and recalculated to obtain the best estimates corresponding to the UN definition of long-term international migrants (NIDI, 2009).

The age structures of the migrants have been reconstructed using the Rogers-Castro model and partial information (six broad age groups) concerning the age structure of migrants to and from Germany (this was the only country for which the disaggregation of international migration by origin, destination and age group was available to us). Thus, the sex and age composition of flows between Germany and all other countries was estimated based on the German data on emigration or immigration, as appropriate. Emigration flows from countries other than Germany to all countries except Germany were assumed to have the same age and sex structure as the immigration flows from this country to Germany.

The regional distribution of migrants was based on the population weight of the destination regions in relation to the total population of each country. Van der Gaag and van Wissen (2002) noted that this method of distributing international migrants is correct if better predictive indicators, such as the historical data on the regional shares of flows, are not available. It is worth noting that the use of the information on foreign population stocks, even if available, instead of the information on total population would not improve the results in the situations where a large proportion of inflows consists of the returning citizens of the destination country (as is the case for example in the flows from Germany to Poland).

International Migration from the Rest of the World

The estimation of net international migration between each of the modelled states and the Rest of the World involved a very substantial uncertainty. For each country, the overall net migration was calculated and then the net migration from the countries within the modelled system was subtracted. The difference between the total net migration and the net intra-system migration was assumed to be the net migration with the Rest of the World. The geographical allocation of net migrants to regions was performed using the same method as in the case of international migration between the states.

[4]*MIMOSA* (*MI*gration *MO*delling for *S*tatistical *A*nalysis) is an abbreviation for the project on 'Modelling of statistical data on migration and migrant populations' funded by Eurostat and conducted in 2007–2009.

Scenarios for the Components of Population Change

In the projection, a population scenario based on the authors' educated guesses of the change in key fertility, mortality and migration indicators was used. The adopted assumptions are discussed below.

Assumptions on Mortality

The mortality scenario has been defined via the changes in life expectancy at birth for males and females. All countries have been divided into four categories based mostly on two variables: economic performance and recent trends in life expectancy. It was assumed that there is a link between economic well-being and changes in age-specific mortality patterns. High income countries with growing economies would have higher life expectancies than low income countries with shrinking economies.

There is plenty of evidence that the changes in life expectancy in Central and Eastern Europe occurred due to shifts in the intensity of mortality in some rather than all the age groups. A decrease in life expectancy observed in some countries, in particular for male populations, was due to the higher mortality in the older working age groups (Okólski, 1987, 1993; Meslé, 1991; Guo, 1993; Hertrich & Meslé, 1999). An increase in life expectancy occurred mainly due to falling infant mortality (Hertrich & Meslé, 1999).

The mortality scenarios were set via two parameters specified for each country and each 5-year projection step: the target life expectancy and the regime of change of age-specific mortality rates. The target life expectancies for post-socialist countries and mature capitalist economies were calculated based on the assumption of either a continuation of the reduction in mortality observed in the last decade or an onset of such a reduction. Two different strategies have been adopted in the modelling of the speed of these changes. For the low mortality countries, it was assumed that the decrease in mortality would slow down over time, as practiced by most national statistical offices (Cruijsen & Eding, 1999) in order to express the belief of experts that the higher the life expectancy, the more difficult a further reduction of mortality would be. The countries with high mortality would increase their gains in life expectancy as their economic situation improved.

In the post-Soviet and Baltic states, a rapid decrease in life expectancy was observed in the 1980s. It was assumed that this trend would be reversed in the post-Soviet group of countries and that a slow improvement would occur. For the Baltic states, the trend had already changed at the time of setting the scenarios. The timing and the strength of the reversal was a factor differentiating the two classes.

Our scenarios were less optimistic than those assumed by the US Census Bureau's International Data Base (IDB) and, in general, ignored the warning that in the past forecasters were over-pessimistic in their mortality decrease prediction (Cruijsen & Eding, 1999; Rees, Kupiszewski, Eyre, Wilson, & Durham, 2001). For Belarus, Moldova and Ukraine, the values assumed were slightly lower than those assumed by Andreev (1995) in his main scenario, but higher than those in his pessimistic one. This relative pessimism was motivated by the belief that economic recovery, the European integration processes and the transformation of lifestyles in Central and Eastern Europe would be slow and painful processes with uncertain success.

Fertility Changes

Based on the arguments provided by Lutz (1996) and Rees (1996b), a fertility scenario built around two qualitative assumptions was adopted. The first, also supported by a large number of other researchers, represented for example by Palomba (1999), was that the changes in lifestyles, values and preferences observed in Europe were permanent phenomena leading to a long-term reduction of fertility. There was and still is controversy as to whether an increase in fertility rates similar to that observed recently in Sweden will be a widespread feature. This was thought to be unlikely to happen in Central and Eastern Europe, as such an increase was attributed to the highly developed social security and maternal benefits system, which would not be affordable to any of the post-socialist countries. Instead, the competition on the labour market and the modernisation of rural areas might have caused a further reduction in fertility.

The second assumption was that there would be a limited convergence in the values of total fertility rates, reducing the gap between the highest and the lowest values from 0.8 observed in 1994 to 0.4 in 2019. The values adopted by us were lower than those in de Beer and van Wissen's (1999) uniformity scenario and similar but not identical to those from the IDB (US Census Bureau, 1999).

Internal Migration Changes

It was assumed that there would be no changes in the intensity of internal migration and that a substantial part of migration-induced population shifts would occur between urban centres and their suburban hinterlands. At the geographical scale in which the projection was conducted the substantial share of these shifts remained intra-regional (Kupiszewski & Rees, 1999).

International Migration Changes

It was assumed that in the 1994–2019 period there would be moderate economic growth in all the countries except Romania, Moldova, Belarus and Ukraine. Simultaneously, migration policies would be tightened by all countries experiencing economic growth. The migration of *Aussiedler* would be slowly reduced to zero due to the exhaustion of potential candidates who could prove German roots. Poland, the Czech Republic, Slovenia, Estonia and Hungary – the countries which at the time were expected to join the EU in 2004, would, as a result, experience a limited increase in migration to/from Austria and Germany and between themselves. It should be noted that Slovakia, Latvia and Lithuania were not seen as serious candidates to the EU at the time. Restrictive migration policies of the EU would limit immigration.

These qualitative assumptions were quantified in a simple way, assuming a reduction of outflows from all countries except post-Soviet and Romania by one third in the first projection period (1994–1999) and an increase of migration between the EU countries (including new members) by 50% after 2004. No changes in the intra-system international migration were expected after 2009, but a 5% reduction in inflows from the Rest of the World was assumed. After 2014, international migration numbers were set to be constant, which reflected our inability to propose a reasonable long-term scenario rather than a belief that migration would really stabilise.

The scenario presented above was rather static, assuming that a temporary reduction in international migration between the modelled states would be partially offset by the admission of the applicant countries to the EU. It was assumed that no rapid changcs would takc placc.

Results: The Future of the Elderly Population of Central and Eastern Europe as Seen in 1995

Under the assumptions described above we investigated the regional change in elderly populations. In this analysis the focus is not on the changes in population numbers but mainly on the changes in the demographic structures. Measures of the advancement of the ageing of the population include the mean age and the percentage of population over a certain age, and dependency ratios are designed to express how many people in a certain age bracket there are per person in another age bracket. In particular, the old age dependency ratio (ODR) is defined here as the number of people over 60 years per 100 population in the working age (20–59 years).

In 1994, the ODR pattern for males showed, in general, a gradual increase from the north-east (below 15 per 100 in the Baltic Republics, Belarus, Northern and Western Poland and the north-eastern part of the former GDR) to the south (above 24 in Southern Romania). For females, the pattern was slightly different: Central and Eastern Europe could be divided into three areas: high ODRs in the south-western belt of Germany and Austria, medium-level ODRs in the north-east (the former Soviet Union countries) and low ODRs in the centre, going from Poland to Romania. However, these areas were by no means homogeneous. For the male population, Romania experienced the largest inter-regional differences with two regions having an ODR in the range 33–36 and two regions having an ODR in the range 15–18. A similar differentiation could be seen in Hungary, Poland and Austria. Germany and the Czech Republic demonstrated a slightly smaller diversity. As expected, there were striking differences in the ODR values between the sexes. Not only were the values of the ODR considerably higher for females than for males, but their spread was also much larger.

After 25 years from the beginning of the projection, the dividing line between the high and the low ODR values basically cuts off Belarus, Moldova and Ukraine from the rest of Central Europe. West of the dividing line, an increase in the ODR for both sexes was predicted to be very considerable. For males, 55 regions would have ODR values above 24 (two regions in 1994). For females, an ODR below 25 would be observed in only one region, in comparison to 152 regions in 1994. The unweighted mean value of the ODR, calculated over all studied regions, would increase from 16 to 23 for males and from 26 to 35 for females. The spread of values for males would increase, becoming comparable to that of females.

To assess the impact of ageing on regional populations, the differences in the ODR at the starting and final points of the projections were analysed. This was done by subtracting the latter from the former and mapping the results. For males, four units: Belarus, Ukraine, Moldova and Arad in Romania would decrease their ODRs. In the three former countries, the very high observed and predicted overmortality in the working age apparently contributed to the phenomenon. The highest increases in the ODR – over 15 per 100 – would be observed in north-eastern Germany. The rest of Germany, the Czech Republic, Slovenia, Austria excluding Steiermark, Slovakia excluding Zapadoslovensky Kraj, Estonia and several voivodships in south-western and western Poland would all have an increase of ODR in the range between 10 and 15. Hungary, Romania, the rest of Poland, Latvia and Lithuania experienced a considerably smaller degree of population ageing.

The distribution of the increase in the ODR for the female population would have a different pattern. Four out of five regions in which the ODR would increase by more than 15 are located in Poland: Warsaw, Łódź, Szczecin and Legnica. The three former are large cities; the fourth is a highly industrialised region. Bratislava would be the fifth region with an extreme increase in population ageing as a direct consequence of past internal migration patterns with a high number of females in their twenties migrating from rural areas to urban centres in 1970s and 1980s (Kupiszewski, Durham, & Rees, 1996). This group would be in their sixties and seventies in 2019. All the Baltic states, most of Poland, the Czech Republic, Slovakia, Slovenia, central Hungary, western Austria and most of Romania would witness a substantial advancement in population ageing. The rest of Central and Eastern Europe would experience a moderate increase in the ODR. As in the case of the male population, there would be a decrease in the ODR in the three former Soviet Republics (Moldova, Ukraine and Belarus).

Over the period 1994–2019, a substantial change in the numbers of the oldest old (85+) would occur. In all German, Austrian, Czech and Slovak regions, in Slovenia and western Hungary, as well as in highly urbanised and industrialised regions in other countries, there would be a very substantial – often more than twofold – increase in the number of very old population, in relative terms higher for males than females, in absolute terms higher for females than for males. In most of Romania, eastern Hungary and in parts of rural Poland, the old male population would remain stagnant or would increase very moderately. In the former Soviet Union, the increase in the number of the oldest old would either stagnate (Baltic states, Belarus) or would decrease (Moldova), sometimes dramatically (Ukraine). These changes would have a very significant impact on the demand for health and care services. The decline in the numbers of the oldest old in some regions, against pan-European trends, should be treated with the utmost attention as a sign of a very poor epidemiological and sanitary situation.

Summarising, the forecast showed clearly that there would be two parallel patterns of the dynamics of elderly population in Central and Eastern Europe. The Central European pattern encompassing all countries except Moldova, Ukraine and Belarus would be characterised by a high level of advancement of the process of population ageing, measured both by the change of the old dependency ratio and the number of the oldest old. The degree of ageing in Central Europe would have a regional dimension and would be different for male and female populations. The very significant increase in the old dependency ratio and in the 85+ populations analysed above should make social security planners think of the allocation of the resources early enough to absorb the ageing population shock. The East European pattern (Belarus, Ukraine, Moldova) would be characterised by high mortality which would effectively prevent the ageing process.

Ex-post Forecast Error

In order to assess the quality of the forecast we have calculated *ex-post* errors for a 10-year time span (1994–2004). The magnitude of the errors of this forecast have been compared with the magnitude of errors for the 1985 Eurostat projection for the period 1985–1995 taken from Rees et al. (2001).

The changes in the population stocks in some countries were corrected by statistical offices after the last round of censuses, by introducing statistical adjustments. To get rid of the effect of these adjustments on the magnitude of forecast errors we have deducted the corrections from 2004 population stocks for seven countries: Austria, the Czech Republic, Lithuania, Moldova, Poland, Romania and the Slovak Republic. For Moldova, which as from 1997 does not include population on the east bank of the Dniester River in its statistics, the adjustment was calculated based on information from the Council of Europe (2004), while for the other countries, except Lithuania, based on NIDI (2004) and Council of Europe (2004). Lithuania introduced post-census adjustments by recalculating its population and emigration numbers (Statistics Lithania, 2006). The assumption was that the error in their measurement of population was solely due to the under-reporting of emigration. According to United Nations (2002), the necessary adjustment was 200,000 over the 12 years between the censuses. In order to take this into account in our assessment of the projection error, we have calculated a rough estimate of the adjustment for the 1994–2004 period.

The forecast errors (Table 3.2), if the post-census corrections were not removed, varied from 19.7% for Moldova, to 0.2% for Austria. The average unweighted error calculated for the absolute values of the country errors was 2.9%. However, the error for Moldova reduced to 1.6% and the average error reduced to 1.2% when calculated in relation to the population without the post-census adjustments. This magnitude of

Table 3.2 Percentage errors of the population forecast for the period 1994–2004

	Forecasted total population ('000s)	Observed total population including statistical adjustments ('000s)	Error of the forecast (%)	Statistical adjustment ('000s)	Observed total population excluding statistical adjustment ('000s)	Error of the forecast (%) after removal of statistical adjustment(%)
Austria	8,154	8,173	−0.23	13.7	8,160	−0.07
Belarus	9,766	9,825	−0.60		9,825	−0.60
Czech Republic	10,304	10,216	0.86	−51.6	10,268	0.36
Estonia	1,365	1,349	1.20		1,349	1.20
Germany	83,950	82,516	1.74		82,516	1.74
Hungary	10,140	101,071	0.33		10,107	0.33
Latvia	2,264	2,313	−2.09		2,313	−2.09
Lithuania	3,619	3,436	5.33	−108.3	3,544	2.11
Moldova	4,314	3,604	19.71	−643.8	4,248	1.57
Poland	39,121	38,182	2.46	−395.6	38,578	1.41
Romania	22,061	21,685	1.74	−558.2	22,243	−0.82
Slovak Republic	5,397	5,382	0.26	−23.8	5,406	−0.18
Slovenia	1,961	1,997	−1.80		1,997	−1.80
Ukraine	45,921	47,271	−2.86		47,271	−2.86
All countries	248,338	246,057	0.93	−1,767.5	247,824	0.21
Average unweighted error			2.94			1.22

Source: Authors' calculations, Eurostat, Council of Europe, NIDI (2004).

error compares favourably with the average unweighted error for the 1985 Eurostat projection for the period 1985–1995, calculated based on data in Rees et al. (2001), which equalled 3.8%.

Evolution of *MULTIPOLES*

As mentioned earlier, we have presented here one of the earliest versions of *MULTIPOLES*. Since its development, the capabilities of the model have been extended following the needs of individual research projects. Among others, the calculation of the labour force size and composition was implemented, initially for the countries and later on for the regions. These calculations are based on the scenarios for labour force activity rates by age and sex. By changing the assumptions on the future age patterns of the activity rates, it is also possible to model the impact of the changes in the retirement age on the size of the labour force. Moreover, a possibility was added to calculate the hypothetical size of replacement migration needed to prevent the decrease of the population or to prevent the increase of a specific dependency ratio (e.g. old age dependency ratio). The new model was used for example in Bijak et al. (2007, 2008).

The most recent development took place within the on-going DEMIFER project, involving further collaboration with Professor Rees. As a response to longer life expectancy, the modelling can be now performed for the age groups up to 100+. The mathematical engine of *MULTIPOLES* was extended with an option to prepare scenarios for migration to/from the rest of the world in terms of emigration rates and immigration numbers, the approach that is considered methodologically better that the one based on net migration numbers. Moreover, it is now possible to set the scenarios in terms of age-specific fertility and mortality rates for each region, instead of total fertility rates and life expectancy for each country. These are important improvements, given the changing age patterns of fertility and mortality. Assumptions concerning the future labour force activity rates can be set either on the regional or country level.

Conclusion

The chapter presents a modification of well established concepts in the field of multiregional population projections. The *MULTIPOLES* model we have created following the ideas of Rees and colleagues allows for a systemic treatment of the population of a large multinational territory, parting with the traditional approach of country-by-country population projections. In particular, it permits a more elegant inclusion of international migration into the process of modelling population change. In the analysis of the results of the scenario based projection, we concentrated on just a fraction of the information available, namely on the size and ageing of the population.

The *MULTIPOLES* model proved to be an effective tool, tested in a number of research projects. The ex-post errors generated in the forecasts using the model are low. Obviously, forecast errors depend not only on the quality of the model but also on the correctness of the input data and the assumptions. Therefore, reliable statistics are crucial for producing meaningful short-term and long-term forecasts.

In 1989, Philip Rees noted that 'the multistate model has proved to be an adaptable beast and is likely to live on into the 1990s' (Rees, 1989). Today, we can add that it is likely to thrive in the twenty-first century, in parallel to the new modelling approaches.

Acknowledgement The authors of this chapter were lucky to work with Professor Rees on models of population dynamics. The *MULTIPOLES* model is an outcome of this collaboration.

References

Andreev, E. (1995). *The last mortality trend in Belarus, Moldova, Russia and Ukraine*. Paper for the task force meeting on alternative population projections in European countries in economic transition, Groningen.

Bijak, J., Kupiszewska, D., & Kupiszewski, M. (2008). Replacement migration revisited: Simulations of the effects of selected population and labor market strategies for the aging Europe, 2002–2052. *Population Research and Policy Review, 27*(3), 321–342.

Bijak, J., & Kupiszewski, M. (2008). Population and labour forecasts for selected European countries: Assumptions and results. In E. Hönekopp & H. Mattila (Eds.), *Permanent or circular migration? Policy choices to address demographic decline and labour shortages in Europe* (pp. 11–33). IOM: Budapest.

Bijak, J., Kupiszewska, D., Kupiszewski, M., & Saczuk, K. (2005). *Impact of international migration on population dynamics and labour force resources in Europe* (CEFMR Working Paper 1/2005). Warsaw: CEFMR.

Bijak, J., Kupiszewska, D., Kupiszewski, M., Saczuk, K., & Kicinger, A. (2007). Population and labour force projections for 27 European countries, 2002–2052: Impact of international migration on population ageing. *European Journal of Population, 23*(1), 1–31.

Council of Europe. (2004). *Recent demographic developments in Europe*. Strasbourg: Council of Europe.

Cruijsen, H., & Eding, H. (1999, May 3–7). *Latest national mortality forecasts in the ECE area.* Joint ECE-Eurostat work session on demographic projections, Perugia.

de Beer J., & van Wissen, L. (Eds.). (1999). *Europe: One continent, different worlds. Population scenarios for the 21st century.* European Studies of Population Vol. 7. Dordrecht: Kluwer Academic Publishers.

Guo, G. (1993). Mortality trends and causes of death – A comparison between Eastern and Western Europe, 1960s-1980s. *European Journal of Population, 9*(3), 287-312.

Herm, A. (2006). Recommendations on international migration statistics and development of data collection at an international level. In M. Poulain, N. Perrin, & A. Singleton (Eds.), *THESIM: Towards Harmonised European Statistics on International Migration* (pp. 77–106). Louvain-la-Neuve: UCL Presses Universitaires de Louvain.

Hertrich, V., & Meslé, F. (1999). *The health crisis in the Baltic countries: A common and varied experience*. Paper presented to the European Population Conference, The Hague.

Kelly, J. (1987). Improving the comparability of international migration statistics. Contributions by the Conference of European Statisticians from 1971 to date. *International Migration Review, 21*(Winter), 1017–1037.

Kupiszewski, M. (1996). Extra-union migration: The East-West perspective. In P. Rees, J. Stillwell, A. Convey, & M. Kupiszewski (Eds.), *Population migration in the European Union* (pp. 11–35). London: Wiley.

Kupiszewski, M. (2001). *Prognoza dynamiki przemian siły roboczej w Europie Środkowo-wschodniej i jej konsekwencje dla procesów integracji europejskiej – podejście regionalne* (Regional forecast of labour force dynamics in Central and Eastern Europe and assessment of its impact on European integration). Raport z grantu Komitetu Badań Naukowych nr. 6P04E05217.

Kupiszewski, M. (2002). *Modelowanie dynamiki przemian ludności w warunkach wzrostu znaczenia migracji międzynarodowych* (Population dynamics modelling under the increasing importance of international migration). Prace Geograficzne 181. Warszawa: IGiPZ PAN.

Kupiszewski, M., Durham, H., & Rees, P. (1996). *Internal migration and regional population dynamics in Europe: Polish case study*. Report prepared for the Council of Europe and for the European Commission. Leeds: School of Geography, University of Leeds.

Kupiszewski, M., & Kupiszewska, D. (1997). *Projection of Central and East European populations – The model, the data and preliminary results* (Working Paper 97/11). Leeds: School of Geography, University of Leeds.

Kupiszewski, M., & Kupiszewska, D. (1999). *Forecasts of regional structures of the elderly populations in Central and Eastern Europe*. Paper presented to the ECE UN conference Status of the Older Population: Prelude to the 21st Century, Sion, 13–15.12.

Kupiszewski, M., & Kupiszewska, D. (2008). International migration component in population dynamics models. In J. Raymer & F. Willekens (Eds.), *International migration in Europe: Data, models and estimates* (pp. 309–327). Chichester: Wiley.

Kupiszewska, D., & Kupiszewski, M. (2005). *A revision of the traditional multiregional model to better capture international migration: The MULTIPOLES model and its applications* (CEFMR Working Paper 10/2005). Warsaw: CEFMR.

Kupiszewska, D., & Nowok, B. (2008). Comparability of statistics on international migration flows in the European Union. In J. Raymer & F. Willekens (Eds.), *International migration in Europe: Data, models and estimates* (pp. 41–71). Chichester: Wiley.

Kupiszewska, D., Kupiszewski, M., Marti, M., & Ródenas, C. (2010). *Possibilities and limitations of comparative quantitative research on international migration flows* (PROMINSTAT Thematic Working Paper No. 4). Last accessed on June 24, 2010, Available from, www.prominstat.eu

Kupiszewski, M., & Rees, P. (1999, May 3–7). *Lessons for the projection of internal migration from studies in ten European countries*. Joint ECE-Eurostat work session on demographic projections, Perugia.

Lutz, W. (1996). Future reproductive behaviour in industrialised countries. In W. Lutz (Ed.), *The future population of the world. What can we assume today?* (pp. 253–277). Laxenburg: IIASA.

Meslé, F. (1991). Mortality in East European countries. *Population, 46*(3), 599–649.

NIDI. (2004). *Final report on data availability and quality of national demographic data series.* Compilation of Long-term National and Regional Population Scenarios for the 10 New EU Member States and 2 EU Accession Countries, Project report, The Hague.

NIDI. (2009). *The MIMOSA model for estimating the origin-destination migration matrix.* Last accessed on December 12, 2009, Available from, http://mimosa.gedap.be/Documents/Explanation_of_MIMOSA_model.pdf.

Nowok, B., Kupiszewska, D., & Poulain, M. (2006). Statistics on international migration flows. In M. Poulain, N. Perrin, & A. Singleton (Eds.), *THESIM: Towards Harmonised European Statistics on International Migration* (pp. 203–231). Louvain-la-Neuve: Presses Universitaires de Louvain.

Okólski, M. (1987). Umieralność mężczyzn w Europie. Wschodniej i w Europie Zachodniej [Male mortality trends in Eastern and Western Europe]. *Studia Demograficzne, 3*(89), 3–28.

Okólski, M. (1993). East-West mortality differentials. In A. Blum & J.-L. Rallu (Eds.), *European population. 2: Demographic dynamics* (pp. 165–189). Montrouge: John Libbey Eurotext.

Palomba, R. (1999, May 3–7). *Values in life and reproductive intentions*. Joint ECE-Eurostat work session on demographic projections, Perugia.

Poulain, M. (1993). Confrontation des statistiques de migration intra-Européenes: Vers plus d'harmonisation. *European Journal of Population, 9*(4), 353–381.

Poulain, M. (1994). La mobilite interne en Europe. Quelles données statistiques? *Espace, Populations, Sociétés, 1*, 13–30.

Poulain, M. (1996). Migration between the countries of the European Union: Current trends. In P. Rees, J. S. C. Stillwell, A. Convey, & M. Kupiszewski (Eds.), *Population migration in the European Union* (pp. 51–66). London: Wiley.

Poulain, M., Debuisson, M., & Eggerickx, T. (1991). *Proposals for the harmonization of European community statistics on international migration*. Louvain-la-Neuve: Institute of Demography, Catholic University of Louvain.

Raymer, J. (2008). Obtaining an overall picture of population movement in the European Union. In J. Raymer & F. Willekens (Eds.), *International migration in Europe. Data, models and estimates* (pp. 209–234). London: Wiley.

Rees, P. (1989). *Old model faces new challenges: A review of the state of the art in multistate population modelling* (Working Paper 531). Leeds: School of Geography, University of Leeds.

Rees, P. (1996a). Projecting the national and regional populations of the European Union using migration information. In P. Rees, J. Stillwell, A. Convey, & M. Kupiszewski (Eds.), *Population migration in the European Union* (pp. 331–364). London: Wiley.

Rees, P. (1996b). *The second demographic transition: What does it mean for the future of Europe's population?* (Working Paper 96/19). Leeds: School of Geography, University of Leeds.

Rees, P., & Kupiszewski, M. (1999). Internal migration: What data are available in Europe? *Journal of Official Statistics, 15*(4), 551–586.

Rees, P., Kupiszewski, M., Eyre, H., Wilson, T., & Durham, H. (2001). *The evaluation of regional population projections for the European Union* (Eurostat Working Paper, Population and Social Conditions 3/2001/E/n 9). Luxembourg: Office for Official Publications of the European Communities.

Rees, P., Stillwell, J., & Convey, A. (1992). *Intra-community migration and its impact on the demographic structure at the regional level* (Working Paper 92/1). Leeds: School of Geography, University of Leeds.

Rogers, A. (1975). *Introduction to multiregional mathematical demography*. New York: Wiley.

Statistic Lithuania. (2006). *Demographic yearbook 2005*. Vilnius: Statistic Lithuania.

United Nations. (2002). *Report on the Millenium Development Goals: a baseline study. Common country assessment for Lithuania*, Vilnius: United Nations.

US Census Bureau. (1999). *International data base*, Washington.

van der Gaag, N., & van Wissen, L. (2002). Modelling regional immigration: Using stocks to predict flows. *European Journal of Population, 18*, 387–409.

Willekens, F., Por, A., & Raquillet, R. (1981). Entropy, multiproportional and quadratic techniques for inferring patterns of migration from aggregate data. In A. Rogers (Ed.), *Advances in multiregional demography* (Research Report RR-81-06). Laxenburg: International Institute for Applied Systems Analysis.

Chapter 4
Modelling with *NEWDSS*: Producing State, Regional and Local Area Population Projections for New South Wales

Tom Wilson

Introduction

State governments in Australia require state, regional and local area population projections for a variety of planning purposes. Used either directly or as inputs to other models, projections inform regional planning strategies, health service planning, transport modelling, educational enrolment projections, water supply infrastructure plans, household and dwelling projections, and assessments of the state's future fiscal position to name a few examples. In New South Wales responsibility for producing the official population projections rests with the Department of Planning's Demography Unit. Existing off-the-shelf projection software is usually too basic or inflexible for the complex projection needs of state governments, so tailor-made projection systems are normally constructed. This chapter describes the modelling system, the New South Wales Demographic Simulation System (*NEWDSS*), created for and used by the Demography Unit to produce the New South Wales 2008 release set of projections (NSW Department of Planning, 2008).

The plan of the chapter is as follows. Initially there is a discussion of the available modelling options and key decisions which must be made in the construction of any population projection system. The criteria which guided the design of *NEWDSS* are then outlined and summaries of its principal features, chosen in light of the design criteria and available modelling options, are presented. The mathematical details of the three component projection models of NEWDSS are set out in three separate sections. The reviewing and adjustment module of the projection system is described before highlights of the New South Wales 2008 release projections are presented and some concluding remarks, including avenues for further research, are given.

T. Wilson (✉)
School of Geography, Planning and Environmental Management, University of Queensland, QLD, Australia
e-mail: tom.wilson@uq.edu.au

J. Stillwell, M. Clarke (eds.), *Population Dynamics and Projection Methods*, Understanding Population Trends and Processes 4, DOI 10.1007/978-90-481-8930-4_4,

Decisions in the Construction of a Population Projection System

Those involved in designing and constructing population projection systems must make many decisions about how populations are to be modelled and how the system will operate. There are many choices but limited advice from the literature on the most appropriate options. Phil Rees is amongst the few who has provided guidance on these issues (especially Rees, 1986, 1994, 1997; from which this section draws extensively), together with demographers such as Smith, Tayman, and Swanson (2001), and Isserman (1993). Decisions about the design of a projection system have important empirical implications. As Isserman has warned, 'Innocent looking technical choices can cause large differences in population projections.' (Isserman, 1993, p. 45). Explicitly listing the important decisions faced by modellers is useful in guiding thinking during the design phase of any projection system, as well as providing a template for succinctly describing the system to users and other modellers. Table 4.1 outlines many of the key issues.

Purpose

Probably the logical starting point in the design of any projection system is to consider its purpose (Rees, 1994): what questions or problems does it need to solve? What projection outputs are required to solve these problems? And of the required outputs, what is most important and should be prioritised given the resources available?

Resources

The resources available to construct the system obviously place constraints on what can be achieved. Important questions include: How big is the budget? How much time is available before the projection system has to be delivered? What staff expertise and time are available? What computing resources (hardware and software) are available to the project?

Age Detail

Age disaggregation is normally either by single years of age or 5 year age groups. Single year of age projections obviously permit the aggregation of results into age groups starting and ending at any single year of age. Education departments in particular are keen on single year of age detail being available because their policy-relevant age groups do not align with standard 5 year age groups. For projections based on the linear rather than the exponential model (see below), the linear approach is less of an approximation for single rather than 5 year age groups (Schoen, 1988, pp. 74–76). For the modeller, however, 5 year age groups make

Table 4.1 Key issues in the construction of a population projection system

Issue	Questions
Purpose	What is the purpose of the projection system? What information is required, and what is most important?
Resources	What budget and expertise are available to construct the system? What is the timeframe like?
Age detail	Will 1 year or 5 year age groups be used? At what age will the final open-ended age group start?
Time detail	Will the projections proceed in 1 year or 5 year intervals? What will the projection horizon be?
Age-time plans	Period-cohort age-time plans are best for most variables; for fertility rates either period-age or period-cohort rates may be chosen
System representation	Macrosimulation, microsimulation or a mic-mac approach?
Geography	Which geographical areas are to be included? How many levels of geography will be modelled?
Internal consistency	If there is more than one level of geography, how will the projections be made consistent between levels? Top down; bottom-up? Will just populations be made consistent, or will demographic components of change also be made consistent?
Exogenous constraints	Will the model incorporate a facility to constrain to external demographic components of change and populations?
Available data	What data are available? How detailed is the spatial and age disaggregation? How long a time series is there? What is the quality like? Are there boundary change problems? Do indirect estimation methods need to be employed to 'fill in' missing data or make adjustments?
Migration concept	Will a movement or transition migration concept be used in the modelling (or both)?
Internal migration model (s)	What sort of internal migration model is best? Fully multiregional, bi-regional approximation, partitioned, migration pool, other directional migration models, net migration model?
Overseas migration model	How will overseas migration be modelled? Immigration flows and emigration flows, immigration flows and emigration intensities, immigration intensities and emigration intensities, net migration? Disaggregation of overseas migration flows into different streams?
Model calculation scheme	Simultaneous, sequential or iterative calculations?
Model specification	Linear or exponential specification?
Programming	Which programming language(s) will be used to operationalise the model?
Formulation of assumptions	What variables will be used to set 'headline' assumptions? What methods will be used to formulate assumptions?
Deterministic or probabilistic	Will the modelling be deterministic or probabilistic?
Projection outputs	What projection outputs are required (linked to the purpose)? How will the outputs be presented? What format(s) do they need to be written out in?
Reviewing outputs	What measures will be used to assess the plausibility and consistency of preliminary projection results?

Sources: Based on the text in Rees (1986, 1994, 1997), plus author's elaborations.

the task much simpler due to considerably less data preparation (less graduation in particular), less expense in purchasing data, fewer assumptions required, and less preliminary projection output to review.

As for the final open-ended age group, a combination of increasing life expectancy, net immigration, and larger birth cohorts reaching the oldest ages is driving rapid population growth at the highest ages. The commonly used age of 85 is now rarely suitable as the start of the final open-ended age group. For the purposes of display in a population pyramid it is helpful if the final open-ended age group contains fewer members than the preceding younger age group. From a modelling perspective a final open-ended age group which is radically changing in age composition within itself is undesirable given that projections of demographic rates for this age group will have been based on quite a different composition. For 5 year age group projections age 90 is probably the lowest reasonable age for the final age group; for a single year of age disaggregation age 100 is probably the lowest reasonable age for most populations.

Time Detail

Projection intervals are normally the same size as age groups. Whilst many users are keen to have projections every year, models which operate in 5 year intervals bring a number of advantages for the producers of projections. Assumptions used in such models do not have to be so concerned about short-term fluctuations in demographic variables, which are very difficult to predict, as *ex-post* evaluations of past projections testify. They are also not subject to assessments of accuracy on an annual basis, which is problematic for at least two reasons. First, accuracy is usually assessed in relation to post-censal population estimates which will inevitably be revised following the next census; and second, assessments of accuracy over just one or 2 years may result in some projection users erroneously concluding that the projections are very inaccurate overall. As for projection horizons, these are often user-defined rather than hard-coded into the model.

Age-Time Plans

Cohort-component projection models, as the name suggests, project the population by cohort over fixed time periods. The age-time plan of survivorship proportions, or rates or probabilities (depending on the model), is therefore period-cohort. (For a description of the various age-time plans, see Rees & Woods, 1986, pp. 305–312). However, initial input data need not be period-cohort. Andrei Rogers's multiregional model, based on a multiregional generalisation of the single region life table, uses period-age data as its initial inputs. Through the multiregional life table period-cohort survivorship proportions are calculated (Rogers, 1995). One type of variable which is an exception to the period-cohort requirement is the age-specific fertility rate used in the projection of births. Many projection models calculate births using period-age fertility rates because fertility data are supplied in this age-time plan. A few models choose to apply period-cohort fertility rates (see for example Kupiszewska & Kupiszewski, 2005).

System Representation

Should the projection system be represented by macrosimulation or microsimulation, or by the more recently invented mic-mac model which includes both types (Willekens, 2005)? Microsimulation is well suited to projections requiring lots of detail on population characteristics (van Imhoff & Post, 1998). The downside of microsimulation models, however, is that they are very data-hungry (often requiring lots of indirect estimation) and require very considerable expertise, time and computing power to construct and implement. Most population projection models are macrosimulation models.

Geography and Internal Consistency

The geographical areas to be included in the model are often those for which projections are required. But this is not always the case. Sometimes the model may be based on smaller 'building bricks' formed by the intersection of boundaries of various regional geographies required by users. Consideration also needs to be given to the question of whether more than one level of geography will be modelled. For sub-national projections it may be useful to project the populations of larger areas (major regions, states and territories, provinces, *et cetera*) which have more reliable and extensive data and constrain the smaller area projections to these larger areas. In the Australian context, there is a strong case for constraining sub-state projections to separate projections prepared at the state level which use data sources for internal and overseas migration unavailable for lower levels of geography. Whether there is a good case for an intermediate level of geography consisting of major regions of a state is more open to debate. The use of indirectly estimated data may strengthen the case for constraining to projections at a higher level of geography based on directly estimated data.

Exogenous Constraints

It is often the case that preliminary summary demographic information is available for the first year or two of a set of projections. For example, by the time census-based Estimated Resident Populations are available for Statistical Local Areas, preliminary births and deaths totals for the first year of the projection horizon may be available for the state as a whole. It is sensible to constraint the projection to these estimates (Simpson, 2004). Alternatively, there may be a desire to constrain regional projections to an independent state-level projection for the entire projection horizon. The modeller needs to decide to what extent external constraints will be incorporated into a projection.

Available Data

The availability and quality of data places some limits on what it is possible to model, particularly in regard to migration data. Is a full origin-destination-age-sex migration matrix available, or is there only a partial matrix? How much age detail is there? Is the quality of data reasonable; if not, how should it be adjusted? Advances in indirect estimation methodology permit unknown sections of data matrices to be estimated in many cases (Willekens, 1999; Raymer, 2007, 2008; Brierley, Forster, McDonald, & Smith, 2008). How close do these indirect estimates get to the real data? Research by Van Imhoff, van der Gaag, van Wissen, and Rees (1997) provides some answers. These researchers conducted a series of empirical experiments to assess which elements of a complete origin-destination-age-sex-time migration matrix could be missing without serious implications in a fully multiregional model. Using iterative proportional fitting, they 'filled in' migration matrices that had some aspects removed or aggregated, and compared the results with the complete matrix. It was found that good approximations of the complete matrix were obtained providing data were available for origin by destination, age by origin, age by destination, and sex by age.

Migration Concept

Will the projection model be based on the movement or transition measure of migration? Movement migration data consist of counts of *migrations* across boundaries over a certain interval of time. Movement data are also known as event data because they are counts of demographic events, like births and deaths. Transition data consist of counts of *migrants* whose locations at the end of a specified interval of time are different from those at the beginning of the interval. Transition data do not include multiple and return moves made during the interval, nor moves made by those individuals who enter or exit the population system during the prescribed time interval. Does it matter which type of migration data the model is based on? 'For those countries where there is a choice of migration data, the choice is not critical as long as the right accounting framework and projection model are matched with the data chosen' (Rees, 1986, p. 76). Discussions of movement and transition migration data are given in Courgeau (1979), Rees (1977, 1985) and Rees and Willekens (1986).

Internal Migration Model(s)

There are a number of options for the internal migration model in terms of the spatial system modelled (Masser, 1976; Rogers, 1976; Wilson & Bell, 2004). Of purely demographic models these include the fully multiregional model, partitioned models, migration pool models, and a bi-regional approximation of the fully multiregional model. Net migration intensity models are best avoided for reasons set out in Rogers (1990). The fully multiregional model works well when applied to just a handful of regions with populations in the millions. As the number of regions increases the amount of data required grows substantially, as does the number of cells in the matrix with zeros or very small numbers. Reduced versions of the fully multiregional model are best considered in these situations (Rees, 1997).

One option is to use a partitioned model. A full origin-destination migration matrix is divided up into a number of parts with limited interactions between the parts (Masser, 1976; Rees, 1996). An example might be the partitioning of a statistical division origin-destination matrix for Australia so that flows between statistical divisions within a state are modelled as origin-destination flows, but flows between statistical divisions in *different* states are handled more simply. Interstate out-migration from statistical divisions in one state might be projected to another state as a whole, and then allocated to statistical divisions in that state using historical proportions.

An alternative is to use a migration pool model (Kupiszewski & Kupiszewska, 2003). This type of model first calculates out-migration from each region (without it being destination-specific); all migrants are effectively placed in a 'pool'; and the pool of migrants is distributed out to destinations using in-migration proportions. Whilst their simplicity and low data requirements are appealing, the age profiles of in-migration proportions in pool models are not intuitively meaningful, difficult to graduate, and difficult to project.

A bi-regional approximation of the fully multiregional model is probably a better option in most cases (Rogers, 1976; Isserman, 1993; Wilson & Bell, 2004). This model has been found to give similar results to the fully multiregional model but requires less detailed data, consequently suffers much less from problems due to small numbers, and is quite simple to implement. Crucially, it maintains the multiregional model's strength of handling directional migration. It also deals with age profiles of in-migration intensities rather than the problematic in-migration proportions of the migration pool model. The model operates by taking each region in turn. Internal migration is projected between the region in question and one large composite region formed by aggregating all other regions. These calculations are repeated for all regions, and a small adjustment made to ensure that internal out-migration summed across all regions equals the sum of internal in-migration.

Overseas Migration Model(s)

Common methods of handling overseas migration in projection models include net migration numbers, emigration intensities and immigration flows, and immigration flows and emigration flows. For a useful review of how overseas migration is handled in population projection models, see Kupiszewki and Kupiszewska (2008). In many models migration flows are implicitly assumed to be homogenous, often because data limitations prevent robust disaggregation. Ideally, however, it would be sensible to divide them up into different streams, either geographically and/or by population characteristics (De Beer, 2008). Different streams often possess separate trends and drivers, and are subject to different migration regulations. For example, because New Zealand citizens have the freedom to enter Australia (unlike other non-Australian nationals) it may be sensible to project their overseas migration movements separately. Similarly, those entering Australia as permanent residents would be separately modelled from non-permanent migrants, such as overseas students and business migrants on fixed-term visas.

Model Calculation Scheme

Rees (1994) distinguishes simultaneous, sequential and iterative calculation schemes for projection models. *Simultaneous* calculations are exemplified by the multiregional matrix model of Rogers (1995): all projection calculations take place in one step. The matrix approach is mathematically elegant and brings insights from the theory of matrix algebra to population dynamics. However, it does not easily permit constraints to be made or checks performed during the development phase of the system, and it requires very large matrices if there are a large number of regions and age groups. *Sequential* approaches proceed step by step. Constraints can be made more easily. *Iterative* calculation schemes are less mathematically tidy, but they provide several advantages. Projection equations do not have to be specially formulated to avoid end-of-interval populations being on the right-hand side of equations, approximations of projection equations sometimes used in sequential calculation schemes can be avoided, and constraints may be easily incorporated. With modern computers the extra time taken to run an iterative model is fairly insignificant.

Model Specification

Should a linear or exponential specification be used? According to the linear specification demographic events are uniformly distributed across the projection interval. Van Imhoff (1990) warns that the linear model can lead can sometimes result in impossible projections and that the assumption of uniformly distributed events is not based on any statistical theory. The exponential model on the other hand, which assumes that demographic rates are constant in the interval, is based on statistical theory and does not lead to impossible results. However, with single year age and time intervals the difference between the linear and exponential models is very small (Schoen, 1988). In addition, the equations of an exponential model are complex (Van Imhoff, 1990). The vast majority of projection models are therefore linear.

Programming Language

Which programming language is best for the operationalisation of a population projection model? The answer is influenced by the purpose of the projection system, its complexity, the projection outputs required, and other decisions made about the projection system. In the 1970s and 1980s the choice of language was limited, and many demographic programs were written in Fortran (for example, Willekens & Rogers, 1978; Rees, 1981). Current projection software uses a variety of programming languages. For example, the *POPGROUP* software package, which is strong in user-friendliness and visualisation, is Excel-based and uses VBA code (Andelin & Simpson, 2005); the family and household projection model *PROFAMY* is written in both Visual C++ and Fortran (Zeng & Wang, no date); the Australian microsimulation model *APPSIM* uses C# (Percival, 2007); and a recently developed regional ethnic group projection model for the UK was created in the statistical package R (Wohland, Rees, Norman, & Boden, 2010). Unfortunately there is almost no publicly available documentation on the programming language choices available to demographic modellers. One of the rare exceptions is Percival (2007). In practice, however, the chosen programming language will often be one familiar to a member of the project team.

Formulation of Assumptions

Whilst considerable agreement exists in the broad approach to population projection modelling (cohort-component methods and, for sub-national projections, multiregional methods), there is no standard set of tools for the formulation of projection assumptions. The simplest approach is to keep recently observed values or trends constant. Alternatively, a qualitative judgement about the trend of a summary indicator – such as the total fertility rate (TFR) or life expectancy at birth - can be made by the modeller. A more rigorous version of this approach is what Wolfgang Lutz describes as an argument-based method, which involves a panel of experts discussing a whole series of influences on future demographic trends (Lutz, 2009; Shaw, 2008).

An alternative approach involves fitting quantitative models to summary indicators or age profiles of demographic rates. A well-known example is the Lee-Carter model and its variants which have proved popular in forecasting life expectancy at birth and age-specific death rates (Booth & Tickle, 2008). An alternative is to calculate many years' worth of age profiles of rates, fit parameterised model schedules, and apply time series models to extrapolate the parameters (Knudsen, McNown, & Rogers, 1993; McNown & Rogers, 1989). These types of models, however, are atheoretical and cannot factor in changes to the determinants of demographic trends. Explanatory models *do* relate demographic trends to predictor variables, but their use in projections is relatively rare. One example is the explanatory model of international migration for Norway created by Brunborg and Cappelen (2009) based on income and unemployment variables. A major challenge in implementing this type of modelling lies in the projection of the predictor variables. Often these are harder to predict than the dependent variable itself.

Many of the methods for preparing projection assumptions were developed – and have been applied – at the national scale. Guidance on the preparation of subnational projection assumptions is rare. Many sub-national fertility and mortality assumptions are linked to a national assumption in a simple manner through the use of fixed regional/national ratios of fertility and mortality (for example, ABS, 2008). A few examples of more complex (but more technically demanding and time consuming) approaches exist. For example, Li and Lee (2005) have adapted the Lee-Carter model to create a coordinated set of mortality forecasts for several populations. De Beer and Deerenberg (2007) created an explanatory model for projecting regional fertility differentials in the Netherlands. For internal migration assumptions, Rees (1997) suggests calculating ratios of regional in- and out-migration rates to the national migration rate and producing linear extrapolations from past time series. Raymer, Bonaguidi, and Valentini (2006) describe an internal migration system in terms of a log-linear model and project the main effects components. Sweeney and Konty (2002) also employ a log-linear model to project internal migration, using VAR models to project the time variables.

Deterministic or Probabilistic

Will the model be deterministic or probabilistic? Most projection models are deterministic: they produce a single set of numbers, or 'point' projections. Probabilistic models attempt to quantify the inherent uncertainty of the demographic future. They produce predictive ranges for projection output variables which indicate the probability of future values lying within those ranges. Probabilistic models overcome some serious shortcomings of the high-low variants approach to demonstrating uncertainty (Keilman, Pham, & Hetland, 2002). However, whilst the methodology is well advanced at the national level, it is in its infancy at the subnational and, in particular, the local scales (Wilson & Bell, 2007; Cameron & Poot, 2009).

Projection Outputs

What outputs will be produced and how will they be presented? The purpose of the model should guide the answer to this question. What derived variables (i.e. those not directly calculated by the projection model) will be needed? These often include growth rates, median ages, sex ratios, and dependency ratios. What kinds of maps are desired? Is three-dimensional visualisation necessary (Bell, Dean, & Blake, 2000)? Will results be made available through a flexible and user-friendly web-based system? In that case a full, detailed database of population accounts of all population stocks and flows will probably be necessary. What format do the output files need to take? This question is particularly relevant if files are to be imported into other software, such as Excel or GIS packages.

Reviewing of Outputs

What checks will be performed to ensure internal consistency and plausibility of the projection outputs? It is quite normal to obtain outputs which require some minor modifications, even with very carefully prepared projection assumptions. This is especially the case at the local area scale where base period data may have undergone considerable smoothing and where data quality may be less than ideal. Do the resulting age profiles look believable in light of migration age profiles and initial cohort sizes? Do sex ratios by age appear within approximately the expected range? Do crude rates of birth, death and migration seem plausible in the light of past trends?

Overview of the New South Wales Demographic Simulation System

Projection Requirements and Desirable Modelling Features

The new projection system was designed to meet the needs of the New South Wales Government. The key criteria were that projection outputs be produced for the state, various sets of official regions and Local Government Areas for at least a 30 year projection horizon (the planning horizon of regional planning strategies), and by age group. These official regions include Department of Planning Regional Strategy areas, Department of Health Health Service Areas, and Department of Ageing, Disability and Home Care Local Planning Areas.

It was also desirable to produce projections for all levels of geography simultaneously in one integrated system. Previously, projections for New South Wales and major regions of the state had been produced first using one model; projections for local areas were prepared later using a completely separate modelling system. Integrating all three levels in one system simplifies the coordination of projection assumptions between geographical levels. Sometimes initial regional assumptions constrain local area assumptions too much, making them less plausible. Fine-tuning of assumptions at the regional level can then provide satisfactory projections at both geographical scales. In addition, the previous local area model projected age-specific migration using net migration rates. Given the conceptual and practical problems with net migration rate projection models (Rogers, 1990), there was a desire to replace this with a directional migration model. Furthermore, from an operational perspective, assumption-setting needed to be fairly simple given the number of spatial units involved. In practice this meant the ability to set 'headline' assumptions in terms of the TFR, life expectancy at birth and net migration.

Key Features of the Projection System

The projection system constructed to meet the above criteria, the *New* South Wales *D*emographic *S*imulation *S*ystem (*NEWDSS*), is outlined in Fig. 4.1, with key characteristics summarised in Table 4.2. Projection assumptions, jump-off populations, look-up tables, various labels for use in output files, and other inputs are gathered in one large Excel workbook. A Fortran 95 program carries out all the computations.

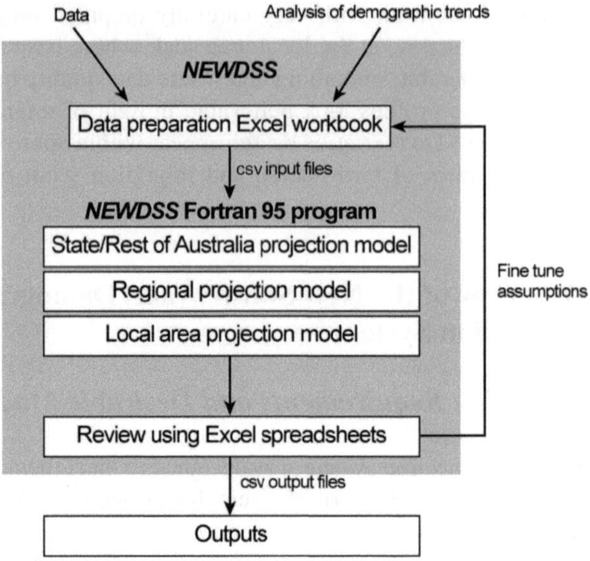

Fig. 4.1 Overview of *NEWDSS*

Input data are provided to the program in the form of comma separated variable (csv) files, which are created by saving relevant sheets from the workbook in csv format. Output files from the fortran program are also written out as csv files so they can be opened automatically in Excel. Initial projection outputs are then subject to a number of checks, usually resulting in some fine-tuning of assumptions for the first few runs of the system.

Table 4.2 Key features of *NEWDSS*

	State/Rest of Australia model	Regional model	Local area model
Age detail	Single years 0–120	Single years 0–120	Five year age groups 0–4 to 80–84 and 85+
Time detail	Single year projection intervals; 50 year max. projection horizon	Single year projection intervals; 30 year max. projection horizon	Five year projection intervals; 30 year max. projection horizon
Age-time plans	Period-cohort for migration and mortality intensities; period-age for fertility	Period-cohort for migration and mortality intensities; period-age for fertility	Period-cohort for migration and mortality intensities; period-age for fertility
System representation	Macrosimulation	Macrosimulation	Macrosimulation
Geography	State and rest of Australia	Between 2 and 20 regions	Between 2 areas per region and a max. of 250

Table 4.2 (continued)

	State/Rest of Australia model	Regional model	Local area model
Internal consistency	Unconstrained	Each accounting equation term is constrained to state projections by sex and age	Total births, deaths and total net migration constrained to regional projections
Exogenous constraints	Constraints to total births, deaths and interstate out-migration permitted (other constraints applied as assumptions)	Constraints to total births and deaths permitted (other constraints applied as assumptions)	Total population constraints applied
Migration concept	Movement	Movement	Transition (but with total population constraints)
Overseas migration model	Emigration rates; No. of immigrations	Emigration rates; No. of immigrations	Emigration probabilities conditional upon survival; No. of immigrants surviving in the country on census night
Interstate migration model	Bi-regional	Bi-regional	Bi-regional approximation of fully multiregional model with all types of internal migration combined into two flows: in-migration; out-migration.
Within-state migration model	n/a	Bi-regional approximation of fully multiregional model	
Model calculation scheme	Iterative	Iterative	Iterative
Model specification	Linear	Linear	Linear
Programming	Fortran 95	Fortran 95	Fortran 95
Deterministic or probabilistic	Deterministic	Deterministic	Deterministic

NEWDSS is a macrosimulation system which models populations at three levels of geography – the state/Rest of Australia, regions, and local areas – with flexibility in the type and number of regions and local areas permitted in order to provide some degree of future-proofing against changes in geographies. In the 2008 release projections, the regional scale comprised 13 major regions, consisting of ABS statistical divisions and subdivisions, whilst the state's 199 Statistical Local Areas (SLAs) were chosen for the local area geography. Projections for statistical divisions are not widely used themselves, with the important exception of Sydney statistical division, but this level of geography is included to provide constraints on the SLA projections. SLA projections are arguably the most important because they comprise the spatial building bricks for Local Government Areas and various sets of official regions. NEWDSS uses look-up tables to aggregate SLAs to these geographies and produce projection output files.

The state/Rest of Australia and regional models use a single year of age breakdown and proceed in single year projection intervals. The period-cohort age-time plan is used for migration and mortality intensities whilst births are projected with period-age fertility rates. The local area model uses 5 year age groups because of the smaller populations involved and produces projections every 5 years. All populations are disaggregated by sex.

The three models are defined primarily by the way in which they handle migration. At the state level a straightforward movement accounts bi-regional model projects interstate migration between New South Wales and the rest of the country. Interstate migration at the regional scale (between each region and the other states and territories) is projected in exactly the same way. For migration between the regions of New South Wales, however, a bi-regional approximation of the fully multiregional model is used.

At the local area scale a transition accounts-based model was constructed because the only directional migration data available for SLAs are transition data collected by the census. It follows Phil Rees's longstanding advice to match your model to the available data (Rees, 1985). It is also useful to match projection intervals with age groups so the local area model was designed to use the 5 year interval migration data (as opposed to the 1 year data, also available from the census). Rather than distinguishing different forms of internal migration, the local area model only handles migration between each local area and the rest of the country in a bi-regional approximation of the fully multiregional model. Trying to distinguish different forms of internal migration, such as interstate migration, inter-regional migration within the state, and inter-local area migration within the region, would be overwhelming given the number of SLAs.

Whilst the core of the local area model is a transition accounts model, it is integrated and consistent with the movement accounts models at the state and regional levels. This is a novel feature of *NEWDSS*. Births, deaths and net total migration as measured from the movement perspective sum across local areas to regional movement accounts births, deaths and net total migration. How this is achieved is described later. Regional projections are also constrained to the state projections.

An iterative approach is used in the computation of all models, following Rees (1981, 1984). Given the speed of modern computers the computing time cost of an iterative approach is small. A run of *NEWDSS* for 13 regions and 199 Statistical Local Areas over a 30 year projection horizon took about 1 minute on a NSW Department of Planning PC in 2008.

Mathematical details of the state/Rest of Australia, Regional and Local Area models are presented in later sections. Notation is set out in Table 4.3. For clarity, time subscripts are generally omitted from flow variables; unless otherwise stated the time period for such variables is always *t*,*t*+1 for the state/Rest of Australia and Regional models and *t*,*t*+5 for the Local Area model.

Table 4.3 Notation

Stocks and flows	
B	Births
D	Deaths
d	Death rate
E	Emigration
e	Emigration rate
ep	Emigration probability conditional upon survival
f	Fertility rate scaled so that rates sum over all ages to unity
GMR	Gross Migraproduction Rate
I	Immigration
i	Proportion of total immigration
M	Interstate migration
m	Interstate migration rate scaled so that rates sum over all cohorts to unity
mp	Migration probability conditional upon survival within the country
NI	Net internal migration (between the area of interest and the rest of Australia)
NIM	Net interstate migration
NPDG	Net private dwelling growth
NTM	Net total migration (all types of internal and overseas migration combined)
NWS	Net within-state migration
OCC	Proportion of dwellings occupied
P	Population
PCE	Population in communal establishments
PD	Private dwellings
PPH	Average number of persons per household
SMR	Standardised Mortality Ratio
TFR	Total fertility rate
WSM	Within-state migration (between regions of the state)
Subscripts and superscripts	
***si*	Population surviving in i at the end of the projection interval
'	Preliminary value
a	Age in single years
A	Age in 5 year age groups
$a \to a+1$	Period-cohort aged *a* at time *t* and aged *a*+1 at time *t*+1
$b \to 0$	Infant period-cohort born during the *t*,*t*+1 projection interval and aged 0 at time *t*+1; or the infant period-cohort born during the *t*,*t*+5 projection interval and aged 0–4 at time *t*+5
*bid**	Population born during the interval who die anywhere before the end
e(RoA)s+	Population existing in the rest of Australia at the start of the interval surviving anywhere in Australia at the end
e(RoA)si	Population existing in the rest of Australia at the start of the interval surviving in *i* at the end
*ei***	Population existing in *i* at the start of the projection interval
*eid**	Population existing in *i* at the start of the interval who die anywhere before the end
eis(RoA)	Population existing in *i* at the start of the interval surviving in the rest of Australia at the end

Table 4.3 (continued)

Stocks and flows	
*eis**	Population existing in *i* at the start of the interval surviving anywhere at the end
eis+	Population existing in *i* at the start of the interval surviving anywhere in Australia at the end
eisO	Population existing in *i* at the start of the interval surviving overseas at the end
eOsi	Population existing overseas at the start of the interval surviving in *i* at the end
f	Females
i	Local area
NSW	New South Wales
OST	Other states and territories combined (i.e. Australia excluding NSW)
r	Region
RoA	Rest of Australia (i.e. Australia minus the area in question)
RoS	Rest of state
s	Sex
t	Point in time
t,t+1	Time interval between *t* and *t*+1 years
t,t+5	Time interval between *t* and *t*+5 years
α	Youngest age of childbearing
β	Oldest age of childbearing

State/Rest of Australia Projection Model

For clarity and brevity description of the state/Rest of Australia model in this section is provided for New South Wales only; however, the Rest of Australia projections are calculated in exactly the same way. Each cohort of New South Wales is projected forward 1 year at a time using the movement accounting equation:

$$\begin{aligned} P^{NSW}_{s,a+1}(t+1) = & P^{NSW}_{s,a}(t) - D^{NSW}_{s,a\rightarrow a+1} - E^{NSW}_{s,a\rightarrow a+1} - M^{NSW\rightarrow OST}_{s,a\rightarrow a+1} \\ & + M^{OST\rightarrow NSW}_{s,a\rightarrow a+1} + I^{NSW}_{s,a\rightarrow a+1} \end{aligned} \tag{4.1}$$

For the cohort born during the projection interval the initial population, $P^{NSW}_{s,a}(t)$, is replaced by the births that occur during the interval, $B^{NSW}_{s,a\rightarrow 0}$. The equations in the rest of this section describe how the accounting equation terms are calculated.

Births by age of mother are projected as the product of period-age age-specific fertility rates scaled to sum to unity across all ages, the TFR and the female person-years at risk:

$$B^{NSW'}_{a} = f^{NSW}_{a}\, TFR^{NSW} \frac{1}{2}\left(P^{NSW}_{f,a}(t) + P^{NSW}_{f,a}(t+1)\right) \tag{4.2}$$

Age-specific fertility rates are decomposed into scaled rates and the TFR to ease alternative scenario design and implementation. It allows different fertility assumptions to be produced by simply altering the TFR but leaving the scaled age-specific rates unchanged, thereby avoiding the time-consuming task of calculating an entirely new set of age-specific fertility rates. Following Simpson (2004), *NEWDSS* also allows projected births to be constrained to a user-supplied total for all ages, B^{NSW}, commonly a recently-published figure for the first year or two of the projection horizon. If this option is taken up then:

$$B_a^{NSW} = B_a^{NSW'} \frac{B^{NSW}}{\sum_{a=\alpha}^{a=\beta} B_a^{NSW'}} \tag{4.3}$$

Births by age of mother are then summed across mothers' age groups and divided into male and female babies according to the sex ratio at birth.

Deaths are projected as the product of period-cohort death rates and the relevant person-years at risk:

$$D_{s,a\to a+1}^{NSW'} = d_{s,a\to a+1}^{NSW} \frac{1}{2} \left(P_{s,a}^{NSW}(t) + P_{s,a+1}^{NSW}(t+1) \right) \tag{4.4}$$

These initial deaths may be constrained to a user-supplied total:

$$D_{s,a\to a+1}^{NSW} = D_{s,a\to a+1}^{NSW'} \frac{D^{NSW}}{\sum_{a\to a+1} \sum_s D_{s,a\to a+1}^{NSW'}} \tag{4.5}$$

Preliminary emigration is calculated by multiplying base period emigration rates by the appropriate person-years at risk:

$$E_{s,a\to a+1}^{NSW'} = e_{s,a\to a+1}^{NSW} \frac{1}{2} \left(P_{s,a}^{NSW}(t) + P_{s,a+1}^{NSW}(t+1) \right) \tag{4.6}$$

Preliminary values are then scaled to all-age and sex emigration totals which are set as projection assumptions:

$$E_{s,a\to a+1}^{NSW} = E_{s,a\to a+1}^{NSW'} \frac{E^{NSW}}{\sum_{a\to a+1} \sum_s E_{s,a\to a+1}^{NSW'}} \, l \tag{4.7}$$

The method of modelling interstate migration differs from many multiregional models. Standard multiregional models are based upon directional migration rates, with net migration being an outcome of the projections. *NEWDSS* takes a modified approach. In order to create a user-friendly method of setting migration assumptions, *NEWDSS* was designed to use assumptions for total (all-age and sex) net interstate migration together with the more usual out-migration rates. Although base period in-migration rates are used in preliminary calculations, final in-migration values are calculated during the projections and are dependent on the net migration and out-migration numbers. This approach allows the headline interstate migration assumptions to be set as net migration whilst retaining the conceptual and practical advantages of modelling directional migration. The details of this approach method are described by Equations (4.8), (4.9), (4.10), (4.11) and (4.12).

Interstate out-migration from NSW to the other states and territories) is projected as the product of period-cohort out-migration rates scaled to sum to unity across all ages, the out-migration Gross Migraproduction Rate and the relevant person-years at risk:

$$M^{NSW\rightarrow OST'}_{s,a\rightarrow a+1} = m^{NSW\rightarrow OST}_{s,a\rightarrow a+1} GMR^{NSW\rightarrow OST}_{s} \frac{1}{2}\left(P^{NSW}_{s,a}(t) + P^{NSW}_{s,a+1}(t+1)\right) \quad (4.8)$$

As with the projection of births, the decomposition of out-migration rates into scaled rates and the GMR is undertaken to ease the changing of assumptions. Out-migration may be constrained to a user-supplied total:

$$M^{NSW\rightarrow OST}_{s,a\rightarrow a+1} = M^{NSW\rightarrow OST'}_{s,a\rightarrow a+1} \frac{M^{NSW\rightarrow OST}}{\sum_{a\rightarrow a+1}\sum_{s} M^{NSW\rightarrow OST'}_{s,a\rightarrow a+1}} l \quad (4.9)$$

Interstate in-migration to NSW is calculated slightly differently. There are three steps. First, preliminary in-migration is calculated using base period in-migration rates:

$$M^{OST\rightarrow NSW'}_{s,a\rightarrow a+1} = m^{OST\rightarrow NSW}_{s,a\rightarrow a+1} \frac{1}{2}\left(P^{OST}_{s,a}(t) + P^{OST}_{s,a+1}(t+1)\right) \quad (4.10)$$

It is not necessary to multiply by the GMR because these preliminary migration numbers are scaled in step 3 below. Second, in-migration is found as total interstate out-migration from Equation (4.9) plus the headline net interstate migration assumption, NIM^{NSW}:

$$M^{OST\rightarrow NSW} = \sum_{a\rightarrow a+1}\sum_{s} M^{NSW\rightarrow OST}_{s,a\rightarrow a+1} + NIM^{NSW} \quad (4.11)$$

In the third and final step the preliminary in-migration calculated in Equation (4.10) is constrained to the result of Equation (4.11):

$$M^{OST\rightarrow NSW}_{s,a\rightarrow a+1} = M^{OST\rightarrow NSW'}_{s,a\rightarrow a+1} \frac{M^{OST\rightarrow NSW}}{\sum_{a\rightarrow a+1}\sum_{s} M^{OST\rightarrow NSW'}_{s,a\rightarrow a+1}} \quad (4.12)$$

Immigration is projected not as rates but as numbers. It is calculated as an immigration proportion (by sex and period-cohort) multiplied by the total immigration assumption:

$$I^{NSW}_{s,a\rightarrow a+1} = i^{NSW}_{s,a\rightarrow a+1} \quad I^{NSW} \quad (4.13)$$

Projecting immigration by decomposing the level from the age pattern permits changes to the overall level of immigration to be made easily without the need to re-calculate age-specific immigration assumptions.

Once the state/Rest of Australia model has completed projections for the whole projection horizon, data for Australia as a whole are calculated by summing the New South Wales and Rest of Australia projections. Some derived projection outputs, such as median age and sex ratios are also calculated. A series of output files is then written out.

Regional Projection Model

At the regional level population projections must not only allow for overseas and interstate migration but also take into account migration within the state. The movement accounting equation for each cohort therefore incorporates additional migration terms for migration to and from each region and the rest of the state :

$$\begin{aligned} P^r_{s,a+1}(t+1) = P^r_{s,a}(t) - D^r_{s,a\to a+1} - E^r_{s,a\to a+1} - M^{r\to RoS}_{s,a\to a+1} \\ -M^{r\to OST}_{s,a\to a+1} + M^{OST\to r}_{s,a\to a+1} + M^{RoS\to r}_{s,a\to a+1} + I^r_{s,a\to a+1} \end{aligned} \quad (4.14)$$

where $P^r_{s,a}(t)$ is replaced by $B^r_{s,b\to 0}(t,t+1)$ for the cohort born during the projection interval. The equations below describe how each of the terms in Equation (4.14) is calculated.

Preliminary births and deaths are calculated in exactly the same way as for the state projections. They are then constrained so that births by age group and deaths by sex and period-cohort sum to the state projections. Regional interstate migration is also modelled as per the state/Rest of Australia model, except for the additional step of constraining to whole-of-state interstate migration.

For migration between the various regions of New South Wales a bi-regional approximation of the fully multiregional model is employed (Rogers, 1976; Wilson & Bell, 2004). There are four steps. First, out-migration from each region to the rest of the state is projected as the product of period-cohort out-migration rates scaled to sum to unity across all ages, the out-migration *GMR* and the relevant person-years at risk:

$$M^{r\to RoS}_{s,a\to a+1} = m^{r\to RoS}_{s,a\to a+1} GMR^{r\to RoS}_s \frac{1}{2}\left(P^r_{s,a}(t) + P^r_{s,a+1}(t+1)\right) \quad (4.15)$$

Second, preliminary in-migration to each region from the rest of the state is calculated as:

$$M^{RoS\to r'}_{s,a\to a+1} = m^{RoS\to r}_{s,a\to a+1} \frac{1}{2}\left(P^{RoS}_{s,a}(t) + P^{RoS}_{s,a+1}(t+1)\right) \quad (4.16)$$

where the rest of the state consists of a different geographical area for each region. These preliminary migration flows do not need to be multiplied by GMRs because they are scaled in step 4. In the third step, total within-state in-migration for each region is calculated as total within-state out-migration plus the headline net within-state migration assumption:

$$M^{RoS\to r} = \sum_{a\to a+1} \sum_s WSM^{r\to RoS}_{s,a\to a+1} + NWS^r. \tag{4.17}$$

In the fourth and final step iterative proportional fitting is used to adjust preliminary in-migration to (i) out-migration by sex and period-cohort calculated in Equation (4.15) (because the number of within-state in-migrations must equal the number of within-state out-migrations):

$$M^{RoS\to r''}_{s,a\to a+1} = M^{RoS\to r'}_{s,a\to a+1} \frac{\sum_r M^{r\to RoS}_{s,a\to a+1}}{\sum_s M^{RoS\to r'}_{s,a\to a+1}} \tag{4.18}$$

and (ii) total within-state in-migration calculated in Equation (4.17):

$$M^{RoS\to r'''}_{s,a\to a+1} = M^{RoS\to r''}_{s,a\to a+1} \frac{M^{RoS\to r}}{\sum_{a\to a+1} \sum_s M^{RoS\to r''}_{s,a\to a+1}}. \tag{4.19}$$

Once convergence has been achieved $M^{RoS\to r'''}_{s,a\to a+1}$ become the final values of within-state in-migration, $M^{RoS\to r}_{s,a\to a+1}$.

Preliminary projections of immigration are calculated by distributing the total immigration assumption for the region by sex and period-cohort:

$$I^{r'}_{s,a\to a+1} = i^r_{s,a\to a+1} I^r. \tag{4.20}$$

The preliminary figures are then iteratively constrained to (i) state immigration by sex and period-cohort:

$$I^{r''}_{s,a\to a+1} = I^{r'}_{s,a\to a+1} \frac{I^{NSW}_{s,a\to a+1}}{\sum_r I^{r'}_{s,a\to a+1}} l \tag{4.21}$$

and (ii) total regional immigration assumptions:

$$I^{r'''}_{s,a\to a+1} = I^{r''}_{s,a\to a+1} \frac{I^r}{\sum_{a\to a+1} \sum_s I^{r''}_{s,a\to a+1}} \tag{4.22}$$

until convergence has been achieved and $I^{r'''}_{s,a\to a+1}$ is assigned to the final regional immigration projection, $I^r_{s,a\to a+1}$.

Preliminary emigration projections are calculated by multiplying base period emigration rates by the appropriate person-years at risk:

$$E^{r'}_{s,a\to a+1} = e^r_{s,a\to a+1} \frac{1}{2} \left(P^r_{s,a}(t) + P^r_{s,a+1}(t)\right). \tag{4.23}$$

Iterative proportional fitting constrains the preliminary values to (i) state emigration by sex and period-cohort:

$$E^{r''}_{s,a\to a+1} = E^{r'}_{s,a\to a+1} \frac{E^{NSW}_{S,a\to a+1}}{\sum_r E^{r'}_{s,a\to a+1}} \tag{4.24}$$

and (ii) total regional emigration assumptions:

$$E^{r'''}_{s,a\to a+1} = E^{r''}_{s,a\to a+1} \frac{E^r}{\sum_{a\to a+1} \sum_s E^{r'}_{s,a\to a+1}} \tag{4.25}$$

No constraining of regional populations to state populations is required because all projected components by sex and period-cohort have been constrained. After *NEWDSS* has calculated derived variables (such as median age and the sex ratio) it writes out a series of output files for the regional level projections.

Local Area Projection Model

Projection calculations at the local area level are more complex than those at the other levels. Whilst the core of the projection approach is the classic transition accounts-based model (Rees & Wilson, 1977; Rees, 1981), total population constraints are applied to each local area using two other models. Which of the two models is applied depends on whether the area is classified as metropolitan (within Sydney statistical division, Newcastle statistical subdivision and Wollongong statistical subdivision) or non-metropolitan (any other part of New South Wales). Note that in this section population terms are assumed to refer to the $t, t+5$ period unless otherwise stated because the projections proceed in 5 year intervals.

Metropolitan Total Populations

Preliminary metropolitan local area total populations are derived from a housing unit method (currently calculated outside *NEWDSS*) because dwelling growth in urban areas is constrained by the existing built environment. The NSW Department of Planning operates a Metropolitan Development Program to monitor and forecast residential land and dwelling developments. The Program maintains a spatially detailed database of forecast net private dwelling increase over the coming 10 years (NSW Department of Planning, 2009). Beyond 10 years dwelling forecasts are made using an expert judgment approach based on a combination of factors, including past dwelling growth, capacity for future dwelling construction, and land release timetables for Sydney's North West and South West Growth Centres.

The projected number of private dwellings *(PD)* is simply the number of private dwellings at the start of the projection interval plus the net private dwelling growth *(NPDG)* over that time period:

$$PD^i\,(t+5) = PD^i\,(t) + NPDG^i \tag{4.26}$$

The projected population is then calculated as the product of the number of private dwellings, the proportion occupied and the average number of persons per household, plus the population in communal establishments:

$$P^{i'} = PD^i \, OCC^i \, PPH^i + PCE^i \tag{4.27}$$

Projections of occupancy, the average number of persons per household and the population in communal establishments are obviously required. Due to data limitations a rather ad hoc approach was taken in the 2008 release projections where these variables were either held constant or extrapolated using either linear or nonlinear functions (depending on what appeared to be reasonable and plausible in light of historical data and qualitative judgement about future developments).

Non-metropolitan Total Populations

Preliminary non-metropolitan local area total populations are calculated via a movement accounting equation including net total migration *(NTM)*, defined as the balance of net migration between the local area in question and the rest of the world:

$$P^{i'}(t+5) = P^i(t) + B^i - D^i + NTM^i \tag{4.28}$$

Net total migration is set as a projection assumption. The reason for aggregating all local area migration into a net total migration term is to simplify data preparation and migration assumption-setting. A full set of base period population accounts, either movement or transition, is difficult to calculate for local areas because emigration data are lacking. Without applying resource-consuming indirect estimation methods, the only way of balancing a population accounting equation for a local area is to use residual net migration, defined as population change over a period minus natural increase. State government demography teams often have access to time series of population accounts for local areas in which growth is divided into births, deaths and net total migration. More detailed historical migration data are often unavailable.

The births and deaths in Equation (4.28) are calculated as follows. Preliminary births by age of mother are found as the product of scaled age-specific fertility rates, the TFR and the female population at risk:

$$B_A^{i'} = f_A^i \, TFR^i \frac{5}{2}\left(P_{f,A}^i(t) + P_{f,A}^i(t+5)\right) \tag{4.29}$$

Constraining is applied to ensure consistency with regional births:

$$B_A^i = B_A^{i'} \frac{\sum_{a,a \in A} B_a^r}{\sum_{i,i \in r} B_A^{i'}} \tag{4.30}$$

Preliminary movement accounts deaths are projected as the product of state mortality rates, a sex-specific Standardised Mortality Ratio (*SMR*) for the area and the relevant period-cohort person-years at risk:

$$D^{i'}_{s,A\rightarrow A+5} = d^{i}_{s,A\rightarrow A+5} SMR^{i}_{s} \frac{5}{2} \left(P^{i}_{s,A}(t) + P^{i}_{s,A+5}(t+5) \right) \quad (4.31)$$

These preliminary deaths are then constrained to regional deaths by sex and period-cohort:

$$D^{i}_{s,A\rightarrow A+5} = D^{i'}_{s,A\rightarrow A+5} \frac{\sum_{a,a\in A} D^{r}_{s,a\rightarrow a+1}}{\sum_{i,i\in r} D^{i'}_{s,A\rightarrow A+5}} \quad (4.32)$$

Both metropolitan and non-metropolitan local area preliminary total populations are scaled to sum to regional projected populations:

$$P^{i} = P^{i'} \frac{\sum_{a\rightarrow a+1} \sum_{s} P^{r}_{s,a\rightarrow a+1}}{\sum_{i,i\in r} P^{i'}} \quad (4.33)$$

These populations form constraints used in the transition-accounts model projections.

Transition Accounts-Based Model

Preliminary local area projections are based on the transition accounting equation:

$$P^{**si'}_{s,A+5}(t+5) = P^{ei**}_{s,A}(t) - P^{eid*}_{s,A\rightarrow A+5} - P^{eisO}_{s,A\rightarrow A+5} - P^{eis(RoA)}_{s,A\rightarrow A+5} + P^{e(RoA)si}_{s,A\rightarrow A+5} + P^{eOsi}_{s,A\rightarrow A+5} \quad (4.34)$$

where $P^{ei**}_{s,A}(t)$ is replaced by $B^{i}_{s,b\rightarrow 0}$ for the cohort born during the 5 year projection interval. Projected local area births are calculated in Equations (4.29) and (4.30). Deaths to individuals living in each local area at the start of the projection interval, $P^{eid*}_{s,A\rightarrow A+5}$, are calculated as:

$$P^{eid*}_{s,A\rightarrow A+5} = P^{ei***}_{s,A}(t) \frac{\left(5d^{i}_{s,A\rightarrow A+5}\right)}{\left[1 + \left(\frac{5}{2} d^{i}_{s,A\rightarrow A+5}\right)\right]} \quad (4.35)$$

whilst for babies born during the projection interval the person-years at risk are changed:

$$P^{bid*}_{s,b\rightarrow 0} = B^{i}_{s} \frac{\left(\frac{5}{2} d^{i}_{s,b\rightarrow 0}\right)}{\left[1 + \left(\frac{5}{2} d^{i}_{s,b\rightarrow 0}\right)\right]} \quad (4.36)$$

This approach to the transition accounts-based model, where all deaths to the start-of-interval population are calculated in one term, is due to Rees (2002). It assumes that irrespective of place of death the origin area's death rates apply. Whilst conceptually inconsistent with the regional projection model, it is insignificant in practical terms.

Amongst the population living in each local area at the start of the projection interval the number alive at the end of the interval is simply the start-of-interval population minus deaths amongst the start-of-interval population:

$$P^{eis*}_{s,A\rightarrow A+5} = P^{ei**}_{s,A} - P^{eid*}_{s,A\rightarrow A+5} \tag{4.37}$$

These survivors are split into emigrants and non-emigrants. Surviving emigrants are calculated as the probability of emigration and survival multiplied by the population at risk (the survivors amongst the start-of-interval population):

$$P^{eisO}_{s,A\rightarrow A+5} = ep^{eisO}_{s,A\rightarrow A+5} P^{eis*}_{s,A\rightarrow A+5} \tag{4.38}$$

Survivors of the start-of-interval population who remain within Australia (+) at the end of the interval are required as the population at risk for internal migration. This term is found by subtracting surviving emigrants from the survivors amongst the start-of-interval population:

$$P^{eis+}_{s,A\rightarrow A+5} = P^{eis*}_{s,A\rightarrow A+5} - P^{eisO}_{s,A\rightarrow A+5} \tag{4.39}$$

The number of people living in local area i at the start of the projection interval who are somewhere else in Australia at the end of the interval is calculated as the product of an out-migration probability and the population at risk (survivors of the start-of-interval population who remain within Australia):

$$P^{eis(RoA)}_{s,A\rightarrow A+5} = mp^{eis(RoA)}_{s,A\rightarrow A+5} P^{eis+}_{s,A\rightarrow A+5} \tag{4.40}$$

The model is a bi-regional approximation of the fully multiregional projection model (Rogers, 1995; Rees, 1997; Wilson & Bell, 2004). It deals with the local area in question (i) and treats the Rest of Australia as the second region. Note that this rest of Australia region possesses a different geographical composition for every local area.

The number of people living in Australia outside local area i at the start of the projection interval who are living in that local area at the end of the interval is found as the product of an in-migration probability and the rest of Australia population at risk:

$$P^{e(RoA)si}_{s,A\rightarrow A+5} = mp^{e(RoA)si}_{s,A\rightarrow A+5} \quad P^{e(RoA)s+}_{s,A\rightarrow A+5} \tag{4.41}$$

Surviving immigrants, $P^{eOsi}_{s,A\rightarrow A+5}$ those *e*xisting *o*verseas at the start of the projection interval who are *s*urviving in Australia in local area i at the end of the interval, are input directly as numbers. These are prepared in advance as part of the projection assumptions.

In the final stage iterative proportional fitting is used to scale the preliminary populations to (i) local area total population constraints:

$$P^{**si'''}_{s,A+5}(t+5) = P^{**si''}_{s,A+5}(t+5)\,\frac{P^{i}(t+5)}{\sum_{A} P^{**si''}_{s,A+5}(t+5)} \tag{4.42}$$

and (ii) regional populations by sex and age group (calculated by the regional projection model):

$$P^{**si'''}_{s,A+5}(t+5) = P^{**si''}_{s,A+5}(t+5)\,\frac{\sum_{a,a\in A} P^{r}_{s,a+5}(t+5)}{\sum_{i,i\in r} P^{**si''}_{s,A+5}(t+5)} \tag{4.43}$$

The outer iterative loop in the local area projection calculations includes all equations except the metropolitan total populations so that all values are successively updated until convergence is achieved.

Reviewing and Adjustment

Once *NEWDSS* has been run projection outputs are subject to a number of plausibility and consistency checks. As a result of this reviewing stage, projection assumptions are usually fine-tuned and the projections re-run. In a complex projections system quite a number of iterations can be expected before obtaining a final set of outputs.

The reviewing and adjustment part of the projections process is far more art than it is science, with no hard and fast rules or quantitative measures of projection plausibility and consistency to rely upon. Judgements about plausibility must be made in light of general knowledge of demographic processes as well as information about local factors which may explain deviations from expected patterns. (This stage of the projections process would certainly benefit from attention by demographers so that a more rigorous and objective approach is devised.) In the New South Wales 2008 release set of projections checks were made of (i) population age profiles, (ii) sex ratios by age, and (iii) aggregate components of change. Figure 4.2 illustrates the spreadsheet used to check SLA age profiles and sex ratios by age.

Population Age Profiles

Many regional and local area populations are characterised by distinctive age profiles, fashioned to a large extent by age-specific migration flows. Many localities undergo only gradual change in their age profiles over time. For example, areas which experience considerable young adult out-migration will often possess an obvious indentation in their age profiles at these ages. If projections indicate a

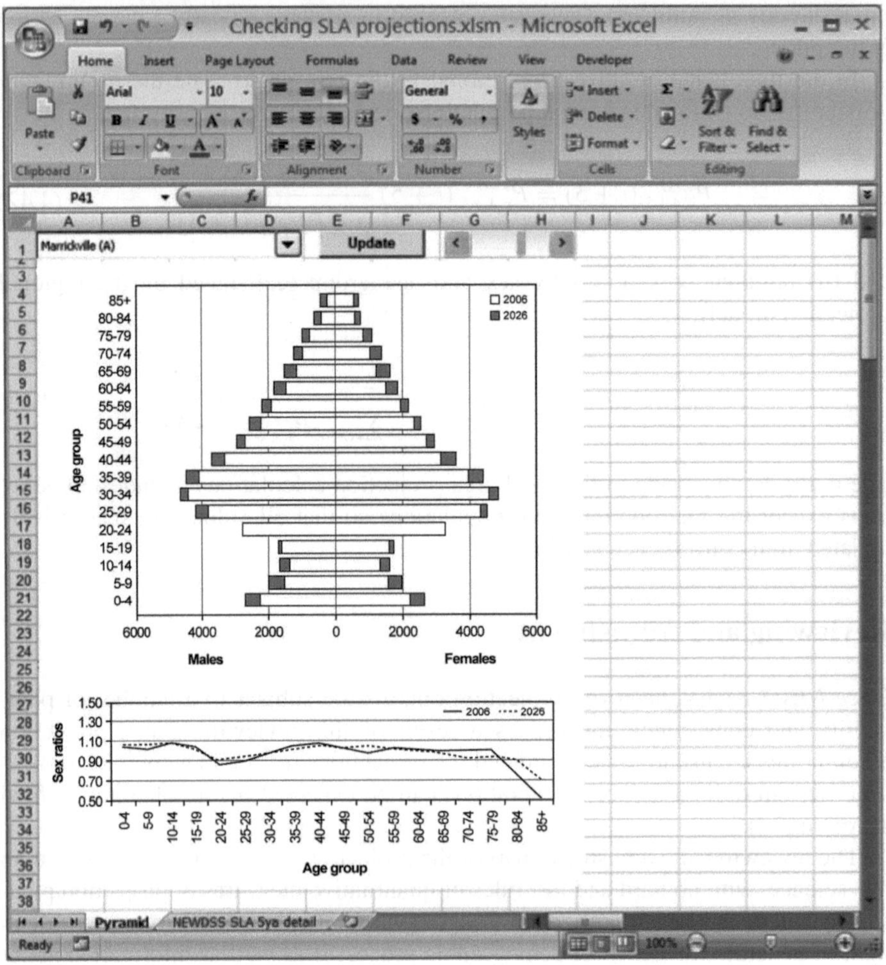

Fig. 4.2 The spreadsheet used to check SLA age profiles and sex ratios by age

loss of this indentation in the future, or a significant shift in its position in the age profile, then there must be a good reason for it; otherwise it is possibly indicative of flaws in the migration age profile assumptions. Similarly, areas which traditionally gain many young adults through migration will tend to feature a protrusion in their age profiles at these ages. Again, if this characteristic changes noticeably in relative size and position in the age profile over time, it implies problems with migration assumptions.

Local areas with significant communal establishment populations, such as prisons and boarding schools, often require adjustments to their migration assumptions in order to maintain plausible projected age profiles. This is due to the migration data on which the assumptions are based failing to fully capture moves into and out of the communal establishment. Communal establishment populations are usually easy to spot in a local area population age profile because they are highly age and sex-concentrated, and create a protrusion in the age profile. Importantly, the age composition of communal establishment populations tends to change little over time. If projections indicate the protrusion in the population age profile changing noticeably over time then adjustments will probably be required.

Sex Ratios by Age

Sex ratios usually change gradually by age. For the youngest children sex ratios reflect the sex ratio at birth of between 105 and 106 male babies per 100 female babies. Unless the net balance of migration is highly sex-selective then the sex ratio of the population will gradually decline with age over the younger and middle adult ages before declining more rapidly in the older adult ages due to higher male mortality. It is quite possible for very slight errors in male and female migration age profiles to compound over time and, several decades into the projection horizon, result in implausible sex ratios in the population over certain ages. For some local areas the age pattern of sex ratios may vary from the 'standard' pattern. Commonly this will be due to communal establishments, but there are also some areas where it is due to certain industries (such as mining) or localised residential patterns.

Aggregate Components of Change in Historical Context

Total births, deaths and net migration over the projection horizon are usefully compared to historical trends. At the local and regional level net migration numbers may well fluctuate considerably over time, but births and deaths tend to change less dramatically. In the absence of any major events (such as the opening of a new prison) it would be expected that projected births, deaths and net migration would broadly follow on from historical values.

Illustration of *NEWDSS*: Projections for New South Wales, 2006–2036

Launch Populations and Projection Assumptions

The projections were launched from the 30th June 2006 Estimated Resident Populations (ERPs) supplied by the ABS. These ERPs are derived from 2006 Census data on the number of usual residents adjusted for (i) net underenumeration (adding in people missed by the census and taking out any double counting), (ii) residents temporarily overseas, and (iii) population change over the short period between 30th June and census night (8th August 2006).

State-Level Assumptions

The TFR for New South Wales is assumed to be 1.85 for the whole projection horizon (apart from a very short smoothing in from recent higher values). This assumption was chosen because (i) at some point the shift of childbearing to older ages will reach biological limits bringing the Cohort Fertility Rate and TFR closer together; (ii) it is assumed that the state (and Australian) economy will remain fundamentally robust, despite the global recession; and (iii) it is believed that government and employers will continue to make progress in enabling both members of a couple to combine work with raising children.

For mortality projections New South Wales sex-specific mortality rates were calculated for age groups 0, 1–4, 5–9, 10–14, ... , 80–84 and 85–89 from 1975–1976 to 2005–2006. Exponential curves were found to provide good fits to past trends and were extrapolated into the future. The 5 year age group rates were then interpolated to single years of age using Beers coefficients (Shryock & Siegel and associates, 1976) and life tables calculated. Life expectancy at birth is projected to rise from 2006–2007 values of 79 years for men and 84 years for women to 86 and 89 years for men and women respectively by 2035–2036. The projections suggest a gradual slowing of life expectancy improvements over the next three decades.

Net interstate migration for New South Wales has fluctuated around –15,000 per annum for the last 30 years, albeit with large deviations associated with changes in the state's economic fortunes relative to the rest of Australia. In recent years, however, net migration has remained heavily negative for about 5 years. Economic factors probably play a major role – the state's unemployment rate relative to the rest of Australia has been rising for about the same period. On the assumption that economic growth will be stronger in the resource-rich states, and incorporating some follow-on effect from higher net overseas migration, net interstate migration is assumed to shift to a more negative long-run level of –20,000 per annum.

Net overseas migration has been set to +50,000 per year, which is at the upper end of the historical range. There were two main reasons for choosing this assumption. Recent Commonwealth Government announcements indicate a policy agenda supportive of high immigration, at least in the short run. In the medium and longer term, changes to the population age structure in the context of continued economic growth are likely to maintain pressure for high immigration. In the next few years, the number of people leaving the labour force age groups will begin to rise as the baby boom generation retires. Providing the economy remains relatively strong, demand for workers is likely to increase.

Regional-Level Assumptions

Regional fertility assumptions were set as TFRs. Regional mortality was projected using the state age schedules of death rates scaled by regional and sex-specific SMRs which were held constant throughout the projection horizon. Regional net migration was first of all set as total net migration (net overseas, interstate and intra-state migration combined) and then disaggregated to separate net overseas, interstate and intra-state streams on the basis of census data covering the last decade.

Local Area Assumptions

SLA assumptions for fertility and mortality were set in exactly the same way as at the regional scale. Migration assumptions were set differently. For SLAs in the metropolitan regions (Sydney statistical division, Newcastle statistical subdivision and Wollongong statistical subdivision) total projected populations are calculated from separate dwelling projections via the housing unit method (as described earlier). The amount of net migration is an output of the projections. For non-metropolitan SLAs net total migration assumptions are formulated for each 5 year projection interval. These assumptions are based on past trends, advice from councils concerning their future economic and population trends (received as part of a consultation of all non-metropolitan councils in NSW in 2008), guidance from the Department of Planning's regional offices, and information on the opening of correctional centres. Small adjustments were made so that net migration assumptions for all SLAs within a region sum to the projected net migration for the region.

A comprehensive discussion of the state and regional projection assumptions and justification for their selection is given in *New South Wales State and Regional Populations Projections, 2006–2036* (NSW Department of Planning, 2008).

The Future Population of New South Wales: Selected Highlights

The 2008 release projections suggest that the demographic future of New South Wales is likely to be one of high population growth, continued population ageing and greater concentration along the coast. Growth is expected to be high relative to most developed countries, with the population of the state increasing from 6.8 million in 2006 to a projected 9.1 million by 2036. Three-fifths of the growth is projected to be due to natural increase with the remainder driven by net overseas migration (New South Wales has long experienced net interstate migration losses). The youthful age structure plays an important role in generating so much natural increase. Population ageing will continue, with the age group 65 years and over increasing its share of the total population from 13½ to 21½% over the 30 years to 2036. The baby boom generation (born 1946–1965) will be instrumental in this transformation.

Figure 4.3 presents a selection of population age structures in 2006 and 2036 for New South Wales and several of its regions. *Sydney*'s age structure reflects its dominant role in Australia's economic and settlement system. It attracts many young adult migrants (from both other parts of Australia and overseas) and sends many older adults off to other parts of Australia (some as returnees) later in their careers or on retirement. *Wollongong*, home to about 280,000 residents is the third largest city in New South Wales after Sydney and Newcastle and is located on the coast just to the south of Sydney. It tends to experience net interstate migration losses and net overseas migration gains at all ages, and a mix of net intra-state migration gains at ages 17 and 18 and net losses in the early 20s, migration patterns

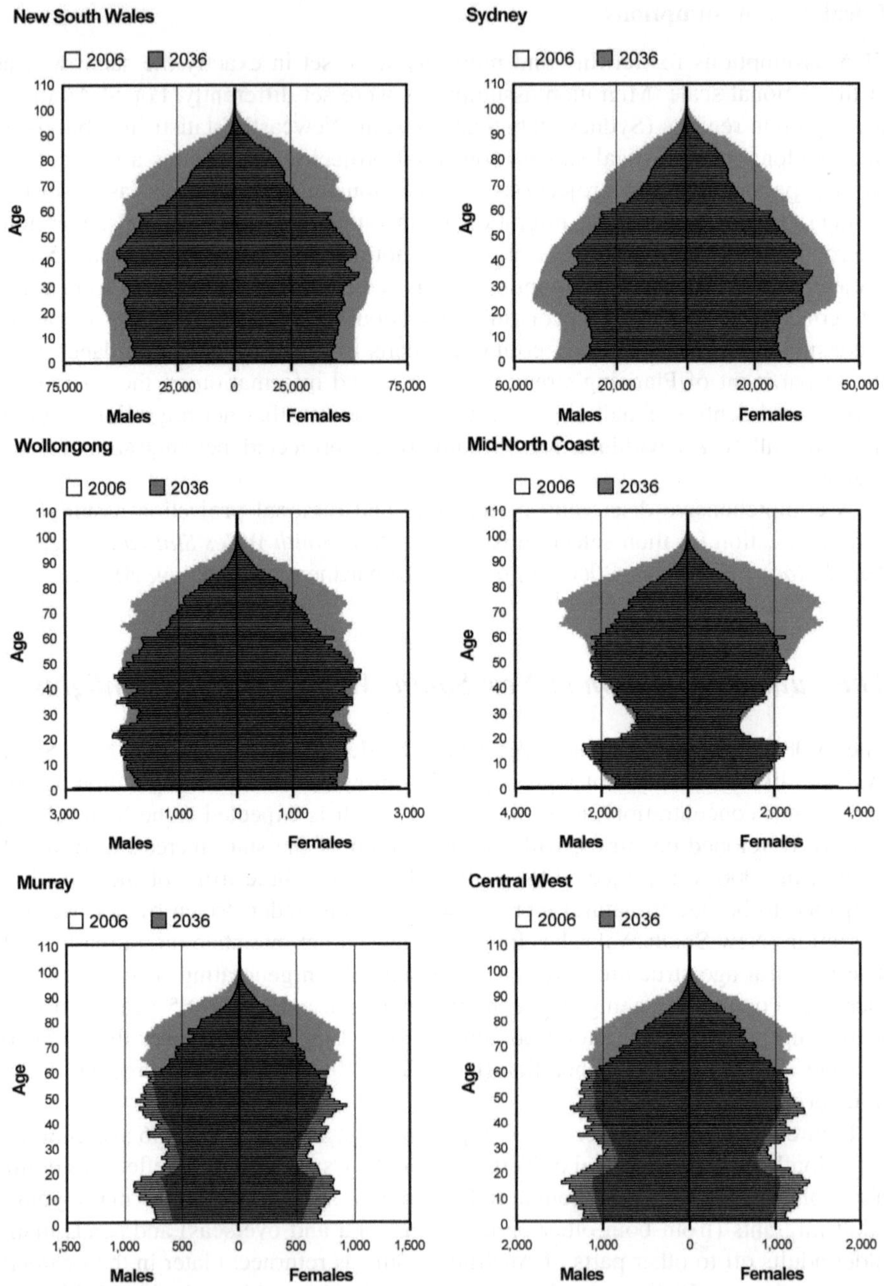

Fig. 4.3 The age-sex profiles of the NSW and selected regional populations in 2006 and 2036 (projected) (Source: 2006 data – Australian Bureau of Statistics; 2036 projections – NSW Department of Planning)

in large part shaped by student moves to and from the University of Wollongong. The *Mid-North Coast* is a scenic region to the north of Sydney with many small towns and a pleasant climate. Its age structure is typical of many non-metropolitan coastal areas of Australia, reflecting young adult out-migration and elderly in-migration. The *Murray* region, extending for about 1,400 km along the Murray River (which forms the border with Victoria) consists mostly of small towns. Its border location and long, thin shape ensure that the bulk of its migration comprises interstate flows to and from Victoria rather than intra-state moves. The net balance of interstate migration is strongly negative in the young adult ages, but positive at older ages, especially around retirement. The region is expected to experience modest population growth in coming decades, along with significant age structure change. The *Central West* is an inland region of NSW which has experienced only modest population growth over the last few decades. Although its total population is not expected to change much over the coming three decades, its age structure certainly will.

Figure 4.4 shows six very different SLA population age structures. *Botany Bay* is a council whose age structure is typical of inner Sydney areas. *Hunter's Hill*, also in inner Sydney, is not typical, being quite a small council area with a large boys' boarding school (which presents quite a challenge in migrating the correct numbers of pupils in and out of the population). *Great Lakes* is a coastal council to the north of Sydney which, quite obviously, experiences substantial elderly in-migration. *Bathurst Regional* is a council area in the Central West region of New South Wales which is divided into two SLAs: part A, which is the regional service city of Bathurst, and part B, which is the rural hinterland. The age profile of part A is influenced noticeably by a campus of Charles Sturt University. *Narromine* is a typical council area in inland New South Wales: its total population has changed little over the last 30 years with natural increase roughly offsetting net out-migration. As its population contracts in the childbearing ages in the future and grows in the elderly age groups natural increase will decline and the total population slowly fall. *Central Darling* presents a more severe case of population decline. It is currently home to about 2,000 residents, and although it is experiencing natural increase this is more than offset by substantial net out-migration.

Considerable spatial variation in population change can be expected across New South Wales in coming decades. Figure 4.5 shows projected percentage population change between 2006 and 2021 by SLA. The boundary between the moderate growth and high growth categories in the map, 16.5%, is the projected state population increase over the period whilst very high growth is classified as two or more times the state population growth. Population increase is projected to be high or very high in the North West and South West Growth Centres of Sydney and a few areas of central Sydney (see the inset map); many areas on, or near, the coast; areas within commuting distance of the Australian Capital Territory (which is represented by the diagonal lines on the map); and many small regional cities. Weak growth or decline is anticipated in the more remote parts of the state, largely reflecting a continuation of past trends.

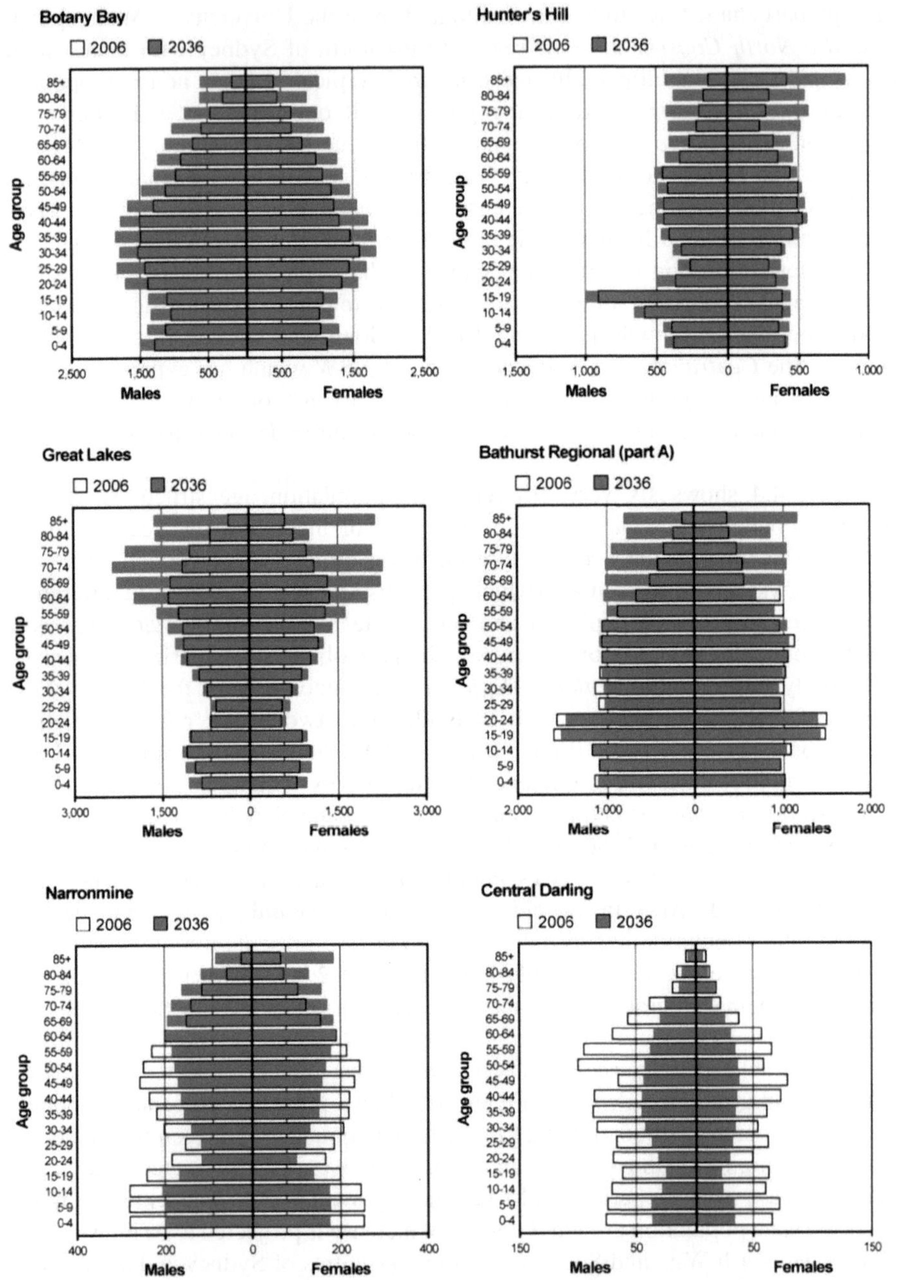

Fig. 4.4 The age-sex profiles of selected SLA populations in 2006 and 2036 (projected) (Source: 2006 data – Australian Bureau of Statistics; 2036 projections – NSW Department of Planning)

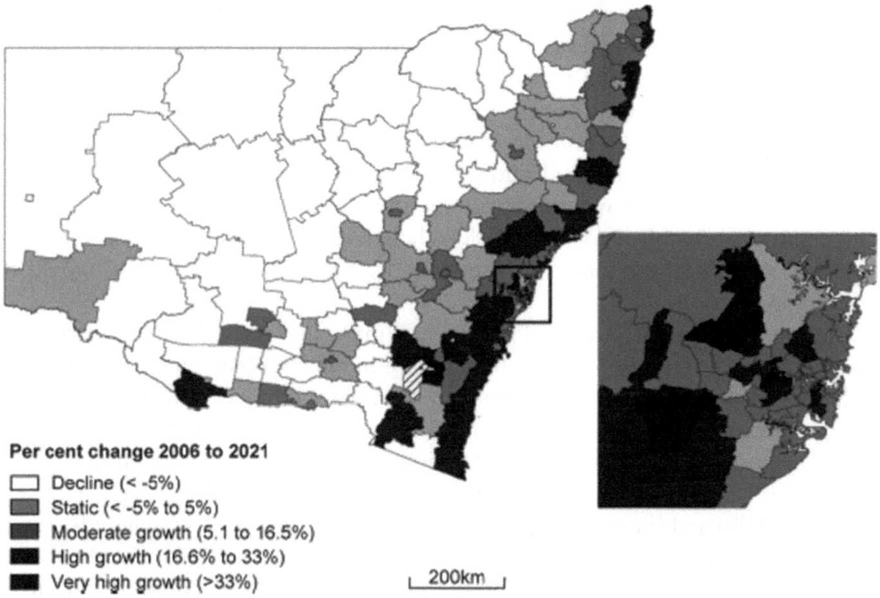

Fig. 4.5 Projected percentage population change in New South Wales by SLA, 2006–2021 (Source: New South Wales Department of Planning, 2008)

Conclusions

This chapter has described *NEWDSS*, the modelling system created and used to produce the 2008 release state, regional and SLA population projections for New South Wales. Highlights of these projections were presented. Two important conceptual advances on existing population projection models in Australia are incorporated into the modelling system. First, movement accounts models to produce state and regional level projections are integrated with a transition accounts model for local areas. Local area population accounts are consistent with the regional level projections in terms of births, deaths and net total migration. The preparation of state, regional and SLA projections in one (rather than two) systems also affords the practical advantage of simplifying coordinated assumption-setting between the three geographical levels. Second, *NEWDSS* projects internal migration at the local area level using age-specific in- and out-migration probabilities rather than net migration rates. The resulting SLA age profiles proved far more plausible compared to the old net migration rate method.

Whilst *NEWDSS* represents an advance on earlier projection systems for New South Wales, there is room for improvement in at least four areas, including: (i) data preparation, (ii) projection modelling, (iii) assumption-setting, and (iv) reviewing preliminary projection output.

In terms of data preparation it would be helpful if there was greater automation in the smoothing of age profiles of rates and probabilities in the data preparation

Excel workbook. Preparation of age-specific fertility and mortality rates is fairly well automated, but this is less the case for migration age profiles.

Regarding the projection modelling, it would be beneficial if the housing unit method for metropolitan SLAs was incorporated into the model, and if the method was also applied to urban SLAs in other parts of the state, where appropriate. The addition of an integer rounding routine prior to output would be helpful. Currently population numbers are written out as real numbers. Scope also remains to deal with communal establishment populations better, especially at the local level. More extensive data checking and error reporting by the projection program would be helpful too.

Current assumption-setting is fairly unsystematic and complicated. In particular, simplification of migration assumption-setting in the data preparation workbook is desirable. Migration assumptions are also too reliant on atheoretical extrapolation and questionable qualitative judgement. They would benefit from being informed more by rigorous analyses of local and regional economic and housing trends, along the lines of those conducted by Stimson, Robson, and Shyy (2009), Baum, Mitchell, and Han (2008) and Freestone, Murphy, and Jenner (2003). Unfortunately few state governments have the resources to conduct such analyses.

The reviewing of preliminary projection results needs to be extended and linked more explicitly to demographic theory and methods. In addition to examining age profiles, sex ratios and aggregate demographic components of change, it might be also be useful to assess projected child/woman ratios against past trends. A greater understanding of how the growth and age structure of local areas respond to changes in local family and household composition, housing market and employment would be beneficial.

It is hoped that at least some of these suggested enhancements will be introduced in the future so that New South Wales demographers will acquire greater benefits from modelling with *NEWDSS*.

Acknowledgement The author gratefully acknowledges the New South Wales Department of Planning for granting permission to include details of the 2008 release population projections for New South Wales.

References

Andelin, J., & Simpson, L. (2005). *POPGROUP population estimation and forecasting system reference manual*. Manchester: Cathie Marsh Centre for Census and Survey Research, University of Manchester.

Australian Bureau of Statistics. (2008). *Population projections, Australia: 2006 to 2101* (Catalogue No. 3222.0). Canberra: ABS.

Baum, S., Mitchell, W., & Han, J. (2008). Socio-economic performance across Australia's non-metropolitan functional economic regions. *Australasian Journal of Regional Studies, 14*(3), 215–249.

Bell, M., Dean, C., & Blake, M. (2000). A model for forecasting the location of fringe urbanisation with GIS and 3D visualization. *Computers, Environment and Urban Systems, 24*(6), 559–581.

Booth, H., & Tickle, L. (2008). *Mortality modelling and forecasting: A review of methods* (Working Paper no. 3). Canberra: Australian Demographic and Social Research Institute, The Australian National University.

Brierley, M. J., Forster, J. J., McDonald, J., & Smith, P. W. F. (2008). Bayesian estimation of migration flows. In J. Raymer & F. Willekens (Eds.), *International migration in Europe: Data, models and estimates* (pp. 149–174). Chichester: Wiley.

Brunborg, H., & Cappelen, A. (2009, September 27–October 2). *An econometric model for forecasting migration to Norway*. Paper prepared for the IUSSP International Population Conference, Marrakech.

Cameron, M. P., & Poot, J. (2009). *Sub-national stochastic population projections with highly volatile international migration: The case of New Zealand*. PAPER prepared for the IUSSP International population conference, Marrakech.

Courgeau, D. (1979). Migrants and migrations. *Population: Selected Papers, 3*, 1–35.

De Beer, J. (2008). Forecasting international migration: Time series projections vs argument-based forecasts. In J. Raymer & F. Willekens (Eds.), *International migration in Europe: Data, models and estimates* (pp. 283–306). Chichester: Wiley.

De Beer, J., & Deerenberg, I. (2007). An explanatory model for projecting regional fertility differences in the Netherlands. *Population Research and Policy Review, 26*, 511–528.

Freestone, R., Murphy, P., & Jenner, A. (2003). The functions of Australian towns, revisited. *Tijdschrift voor Economische en Sociale Geografie, 94*(2), 188–204.

Isserman, A. M. (1993). The right people, the right rates: Making population estimates and forecasts with an interregional cohort-component model. *Journal of the American Planning Association, 59*(1), 45–64.

Keilman, N., Pham, D. Q., & Hetland, A. (2002). Why population forecasts should be probabilistic – Illustrated by the case of Norway. *Demographic Research, 6*(15). (pp. 409–454). www.demographic-research.org

Knudsen, C., McNown, R., & Rogers, A. (1993). Forecasting fertility: An application of time series methods to parameterised model schedules. *Social Science Research, 22*, 1–23.

Kupiszewska, D., & Kupiszewski, M. (2005). *A revision of the traditional multiregional model to better capture international migration: The MULTIPOLES model and its application* (Working Paper 10/2005) Warsaw: Central European Forum for Migration Research.

Kupiszewki, M., & Kupiszewska, D. (2008). International migration component in population dynamics models. In J. Raymer & F. Willekens (Eds.), *International migration in Europe: Data, models and estimates* (pp. 309–327). Chichester: Wiley.

Kupiszewski, M., & Kupiszewska, D. (2003). *Internal migration component in subnational population projections in member states of the European Union* (Working Paper 2/2003). Warsaw: Central European Forum for Migration Research.

Li, N., & Lee, R. (2005). Coherent mortality forecasts for a group of populations: An extension of the Lee-Carter method. *Demography, 42*(3), 575–594.

Lutz, W. (2009). *Toward a systematic, argument-based approach to defining assumptions for population projections* (Interim Report IR-09-037). Laxenburg: International Institute for Applied Systems Analysis.

Masser, I. (1976). The design of spatial systems for internal migration analysis. *Regional Studies, 10*(1), 39–52.

McNown, R., & Rogers, A. (1989). Forecasting mortality: A parameterised time series approach. *Demography, 26*(4), 645–660.

NSW Department of Planning. (2008). *New South Wales state and regional population projections, 2006–2036: 2008 release*. Sydney: NSW Department of Planning.

NSW Department of Planning. (2009). *MDP 2007/08 report*. Sydney: Department of Planning.

Percival, R. (2007). *APPSIM – Software selection and data structures* (Working Paper no. 3). Canberra: National Centre for Social and Economic Modelling, University of Canberra. Accessed on December 7, 2009, from http://www.canberra.edu.au/centres/natsem/publications

Raymer, J. (2007). The estimation of international migration flows: A general technique focused on the origin-destination association structure. *Environment and Planning A, 39*, 985–995.

Raymer, J. (2008). Obtaining an overall picture of population movement in the European Union. In J. Raymer & F. Willekens (Eds.), *International migration in Europe: Data, models and estimates* (pp. 209–234). Chichester: Wiley.

Raymer, J., Bonaguidi, A., & Valentini, A. (2006). Describing and projecting the age and spatial structures of interregional migration in Italy. *Population, Space and Place, 12*, 371–388.

Rees, P. (1977). The measurement of migration, from census data and other sources. *Environment and Planning A, 9*, 247–272.

Rees, P. (1981). *Accounts based models for multiregional population analysis: Methods, program and users' manual* (Working Paper 295/Computer Manual 13). Leeds: School of Geography, University of Leeds.

Rees, P. (1984). *Spatial population analysis using movement data and accounting methods: Theory, models, the 'MOVE' program and examples* (Working Paper 404). Leeds: School of Geography, University of Leeds.

Rees, P. (1985). Does it really matter which migration data you use in a population model? In G. A. van der Knapp & P. E. White (Eds.), *Contemporary studies of migration* (pp. 55–77). Norwich: GeoBooks.

Rees, P. (1986). Choices in the construction of regional population projections. In R. Woods & P. Rees (Eds.), *Population structures and models: Developments in spatial demography* (pp. 126–159). London: Allen and Unwin.

Rees, P. (1994). *Key issues in subnational projection models* (Working Paper 94/8). Leeds: School of Geography, University of Leeds.

Rees, P. (1996, September). *In-round-out: The organisation of information in population projections*. Paper prepared for the First International Geocomputation Conference, Leeds.

Rees, P. (1997). Problems and solutions in forecasting geographical populations. *Journal of the Australian Population Association, 14*(2), 145–166.

Rees, P. (2002). New models for projecting UK populations by ethnic group at national and subnational scales. In J. Haskey (Ed.), *Population projections by ethnic group: A feasibility study* (pp. 27–42). London: Office for National Statistics.

Rees, P., & Willekens, F. (1986). Data and accounts. In A. Rogers & F. J. Willekens (Eds.), *Migration and settlement: A multiregional comparative study* (pp. 19–58). Dordrecht: D. Reidel.

Rees, P., & Wilson, A. G. (1977). *Spatial population analysis*. London: Edward Arnold.

Rees, P., & Woods, R. (1986). Demographic estimation: Problems, methods and examples. In R. Woods & P. Rees (Eds.), *Populations structures and models* (pp. 301–343). London: Allen and Unwin.

Rogers, A. (1976). Shrinking large-scale population-projection models by aggregation and decomposition. *Environment and Planning A, 8*, 515–541.

Rogers, A. (1990). Requiem for the net migrant. *Geographical Analysis, 22*, 283–300.

Rogers, A. (1995). *Multiregional demography: Principles, methods and extensions*. Chichester: Wiley.

Schoen, R. (1988). *Modeling multigroup populations*. New York: Plenum Press.

Shaw, C. (2008). The national population projections expert advisory group: Results from a questionnaire about future trends in fertility, mortality and migration. *Population Trends, 134*, 42–53.

Shryock, H. S., Siegel, J. S. and associates (1976). *The methods and materials of demography: Condensed edition by Stockwell, E. F.* New York: Academic Press.

Simpson, L. (2004). Integrating estimates within population forecasts. *Applied Population and Policy, 1*(2), 89–104.

Smith, S. K., Tayman, J., & Swanson, D. A. (2001). *State and local population projections: Methodology and analysis*. New York: Kluwer Academic.

Stimson, R. J., Robson, A., & Shyy, T. (2009). Modeling regional endogenous growth: An application to the non-metropolitan regions of Australia. *Annals of Regional Science, 43*, 379–398.

Sweeney, S. H., & Konty, K. J. (2002). Population forecasting with nonstationary multiregional growth matrices. *Geographical Analysis, 34*(4), 289–311.

Van Imhoff, E. (1990). The exponential multidimensional demographic projection model. *Mathematical Population Studies, 2*(3), 171–182.

Van Imhoff, E., & Post, W. (1998). Microsimulation methods for population projection. *Population: An English Selection, 10*(1), 97–138.

Van Imhoff, E., van der Gaag, N., van Wissen, L., & Rees, P. (1997). The selection of internal migration models for European regions. *International Journal of Population Geography, 3*, 137–159.

Willekens, F. (1999). Modeling approaches to the indirect estimation of migration flows: From entropy to EM. *Mathematical Population Studies, 7*(3), 239–278.

Willekens, F. (2005). Biographic forecasting: Bridging the micro-macro gap in population forecasting. *New Zealand Population Review, 31*(1), 77–124.

Willekens, F., & Rogers, A. (1978). *Spatial population analysis: Methods and computer programs* (Research report RR-78-18). Laxenburg: Institute for Applied Systems Analysis.

Wilson, T., & Bell, M. (2004). Comparative empirical evaluations of internal migration models in subnational population projections. *Journal of Population Research, 21*(2), 127–160.

Wilson, T., & Bell, M. (2007). Probabilistic regional population forecasts: The example of Queensland, Australia. *Geographical Analysis, 39*, 1–25.

Wohland, P., Rees, P., Norman, P., & Boden, P. (2010). *What happens when international migrants settle? A projection of the UK's ethnic group populations – model, software, components and results* (Working Paper). Leeds: School of Geography, University of Leeds.

Zeng, Y., & Wang, Z. (no date). *ProFamy – Software for household forecasting: User's guide*. Accessed on November 4, 2009, from http://www.sabresys.com/whitepapers/report_no_1.pdf

Chapter 5
Relationships Between UK Sub-national Trends in Infant Mortality and Fertility

Paul Norman

Introduction

If there are more births than deaths in an area then, due to this positive natural change gain, the area's population will grow; assuming a zero impact of net migration. To determine the relative contributions of each demographic component on population change, a variety of models can be used (Woods & Rees, 1986). Dating back to Stone (1971), but developed and promoted by Rees and colleagues (Rees & Wilson, 1973, 1977; Rees & Convey, 1984; Rees, 1986), a 'population accounts' framework ensures that every component of change is explicitly included. A useful model for summarising results is Webb's (1963) population change typology which has been used effectively by Rees and colleagues in a series of influential European reports (see, for example, García Coll & Stillwell, 2000; Kupiszewski, Durham, & Rees, 1996).

A more long-term model, the 'demographic transition model', with its origins in Thompson (1929) and Notestein (1945), illustrates how differential birth and death rates impact on population change. A distinctive feature of the demographic transition model is that mortality decline is followed after a time by reduced fertility (Woods, 1986a). Whilst there is a consensus on the reasons for mortality decline (for example, improved agricultural yields and nutrition, better personal hygiene, the introduction of preventive medicine and public health, reductions in infectious diseases), explanations for why fertility declined are less clear. Lee (2003) notes that economic theories of fertility propose that couples wish to have a number of surviving children and that if potential parents recognise an exogenous improvement in child survival, then fertility will decline.

The aim of this chapter is to add to our understanding of the relationship between different demographic components since trajectories over time combine to affect changes in population size. Clarifying whether changes in infant mortality rates relate geographically with changes in fertility rates will inform both the fertility rate

P. Norman (✉)
School of Geography, University of Leeds, Leeds LS2 9JT, UK
e-mail: p.d.norman@leeds.ac.uk

J. Stillwell, M. Clarke (eds.), *Population Dynamics and Projection Methods*,
Understanding Population Trends and Processes 4, DOI 10.1007/978-90-481-8930-4_5,

assumptions of projection models and the survival of live births through the first year of life. In the next section of this chapter the findings are reported of a previous study on timing and patterns of sub-national variations in infant mortality trends (as child mortality) in relation to fertility trends in England and Wales during the late nineteenth and early twentieth centuries. Then, more recent trends are investigated for 1981–2006 with the study area extended to cover the whole of the United Kingdom. In the section which follows regression models are developed to determine socio-demographic influences on infant mortality and fertility rates. Finally, a classification is developed to highlight whether groups of areas have similar trends in infant mortality and fertility rates.

Regional Variations in Child Mortality and Fertility During the Demographic Transition in England and Wales

An extensive study of sub-national trends in child mortality and fertility was presented in 1977 by William Brass at a joint conference of the British Society of Population Studies and the Population Geography Study Group of the Institute of British Geographers on 'Regional Demographic Development'. A volume edited by John Hobcraft and Phil Rees resulted from this conference including a chapter on this topic by Brass and Kabir (1979). The focus of this work was an evaluation of the evidence of the effects of falling child mortality on fertility. Based on an extensive literature, largely from the medical profession (CICRED, 1975), Brass and Kabir's expectation was that reductions in child mortality had a 'major impact in the establishment of fertility falls' (p. 71).

Using rates of child mortality (deaths under 5 years of age) and general fertility rate, the time period of Brass and Kabir's study was from the 1860s to the 1930s across sub-national regions of England and Wales. The analytical approach was not to investigate the association between levels of child mortality and fertility but between trends of levels over time. The percentage changes in 5-year intervals were calculated (from non-overlapping moving averages) and then correlated. The question asked by the authors was whether movements over time intervals in the measures tend to be related. The analysis was carried out using measures at the same time point, but importantly in this application, with time lags.

To set the scene, Fig. 5.1 illustrates demographic rates relevant to both the demographic transition and to Brass and Kabir's study. From around 1875, crude birth rates are falling in parallel to crude death rates. The general trends are interrupted by the First World War after which, during the 1920s, the decline in death rates appears to steady but birth rates continue to fall. The stepped appearance of infant mortality during the nineteenth century is because data were not available annually until the twentieth century. From 1900 onwards, infant mortality rates fall, largely due to improved living conditions, diet and sanitation (Woodroffe, Glickman, Barker, & Power, 1993) and the increasing influence of pioneers in public and child health such as George Newman (Dunn, 2005; Newman, 1906; Shelton, 2006).

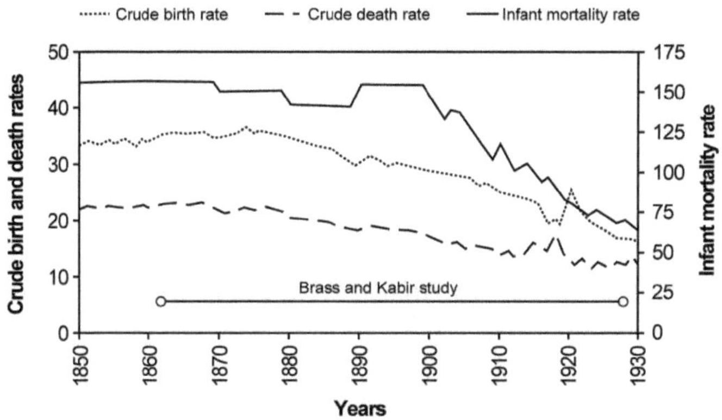

Fig. 5.1 Demographic rates in England and Wales, 1850–1930 (Source: Author's calculations based on data supplied by ONS)

Brass and Kabir (1979) reported the average correlations during their study period with annual time lags within 5-year periods. Along with moving averages from data overlaps, this was to reduce annual fluctuations and to display patterns more clearly. In Fig. 5.2, a time displacement of zero means that the rates were correlated for the same time period. The time displacement is positive when the fertility change is later than the mortality change. If fertility reduction occurred following mortality reduction, then correlations would increase with increasingly positive time displacement. As can clearly be seen in Fig. 5.2, this is not the case since the correlations are highest with little time displacement and lower correlations with increasing time displacement. Brass and Kabir also found this for longer-term time displacements (to over 20 years).

Since the maximum correlation was found at effectively no time displacement, Brass and Kabir (1979, p. 86) concluded there was no detectable 'direct influence of child mortality on fertility' and that time-localised correlations of trends in areas were 'entirely consistent with the view that causative factors were largely the same

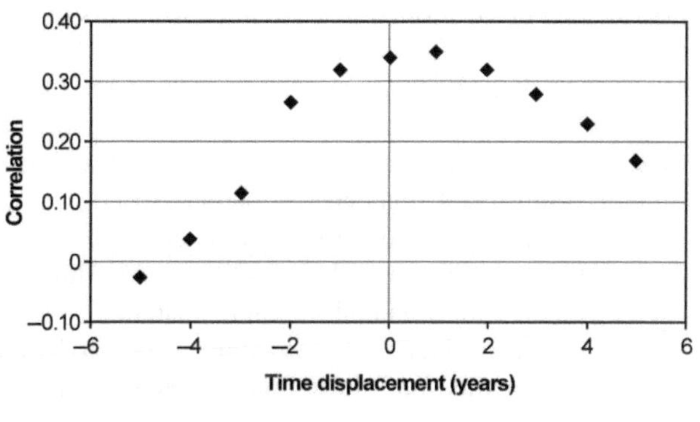

Fig. 5.2 Average correlations between changes in child mortality and general fertility, 1860s–1930s (Source: Brass & Kabir, 1979, table 3.8, p. 81)

for fertility as for child mortality'. However, the authors note temporal and spatial *variations* in declines in child mortality due to variations in the type and pace of socio-economic change but *consistency* in the falls in fertility by region. Given the correlations between child mortality and fertility trends, Brass and Kabir struggle to explain this paradox.

Sub-national Variations in Infant Mortality and Fertility During 1981–2006 in the UK

The demographic transition model relates to the historical decline in mortality and fertility in various European countries and currently in many developing countries (Lesthaeghe, 2005). The recognition that differences in living arrangements occur as social attitudes change and that movements of people between countries lead to populations becoming multi-cultural have led to second (Rees, 1997; van de Kaa, 1987) and third demographic transition models being proposed (Coleman, 2006). The demographic and societal characteristics of the second and third demographic transitions such as low marriage rates, rises in divorce and cohabitation rates, a multi-national, multi-ethnic population experiencing relative and absolute ageing due to very low fertility and general and infant mortality rates are all occurring within the UK (Dunnell, 2007).

Whilst infant mortality rates declined substantially during the twentieth century, infant mortality remains high on academic (Garrett, Galley, Shelton, & Woods, 2006) and public health and policy agendas (Freemantle & Read, 2008) within the UK and elsewhere (Storeygard, Balk, Levy, & Deane, 2008). Since the nineteenth century, the strong tendency has been for infant mortality rates in urban and mining areas to be higher than those in more rural locations, largely due to adverse living conditions and housing density (Guildea, Fone, Dunstan, & Cartlidge, 2005; Lee, 1991). Although to some extent explained by distributions of people by social class, geographic variations in infant mortality have been observed (Botting & Macfarlane, 1990; OPCS, 1978). Over time, the most deprived areas within the UK have had the highest infant mortality rates and the least deprived areas the lowest (Carstairs, 1981; Fitzpatrick & Cooper, 2001; Kmietowicz, 2001). Consistent with Brass and Kabir (1979), Lee (1991) found that the improvements in infant mortality rates continued to be geographically uneven over time (1861–1971). In bringing this time series more up-to-date (1971–2006), Norman, Gregory, Dorling, and Baker (2008a) found that whilst infant mortality rates improved overall, not all locations or area types experienced the same amount of improvement with the strong relationship between infant mortality and area deprivation persisting.

The annual number of births fluctuated widely in the UK's countries during the twentieth century (Chamberlain & Gill, 2005). Three main peaks occurred; one after each of the World Wars and the more sustained 'baby boom' period which peaked in 1964. A 'baby bust' followed during the 1970s and, whilst there was a slight recovery of fertility rates during the 1980s when the baby boom generation reached childbearing age (an 'echo' of the original boom), fertility rates continued to decline to a record low of 1.63 children per woman in 2001 (Dunnell, 2007). Subsequently, fertility rates have risen in all four countries of the UK. During the last 25 years, total fertility rates for Northern Ireland have remained higher than for the UK, but the difference has narrowed. Since the mid-1980s, fertility rates for Scotland have been lower than the rest of the UK with rates for Wales higher than for England during the 1990s, but similar since 2001. Sub-nationally, the fertility decline experienced nationally during the latter part of the twentieth century and the increase in fertility since 2001 are also evident for regions. Fertility trends at local authority level are, however, much more wide ranging and can differ substantially from regional and national trends (Tromans, Natamba, Jefferies, & Norman, 2008).

A large literature investigates why fertility rates change over time. Low fertility and postponement of first birth are associated with increased education and the career aspirations and economic independence of women (Rendall & Smallwood, 2003; Simpson, 2009; Smith & Ratcliffe, 2009). Changing attitudes in society have led to lower marriage rates, thus reducing the expectation that people will have children and women's childbearing intentions (Chamberlain & Gill, 2005). Sub-UK and sub-national variations occur. Country-specific factors such as the cost and availability of housing and variations in the labour market can affect fertility levels. Local factors such as concentrations of people by social class or ethnic group and the presence of students and armed forces bases will influence fertility (Tromans, Natamba, Jefferies, & Norman, 2008). With both national and local impacts, the recent upturn in fertility relates in part to increases in immigration with women born outside the UK found to have higher fertility rates (Tromans, Jefferies, & Natamba, 2009).

The need for a geographical perspective on our understanding of fertility has been identified since 'spatial variations (or the lack of them) can be a useful test of the comprehensiveness of grand theories of population change' (Boyle, 2003, p. 622). Subsequent research on local fertility levels and trends during the latter part of the twentieth and early twenty-first centuries does indeed find distinct geographies to fertility within England and Wales (Tromans, Natamba, Jefferies, & Norman, 2008) and in Scotland (Boyle, Graham, & Feng, 2007). Since sub-national trends in the declines in infant mortality during the same period are also shown to vary geographically (Norman, Gregory, Dorling, & Baker, 2008a), here using UK-wide data for local authorities, Brass and Kabir's (1979) study framework is applied to the period 1981–2006.

The average correlations between changes in infant mortality rate and changes in total fertility rates are illustrated in Fig. 5.3. There are very weak correlations between infant mortality and fertility, whatever the time displacements between the measures. This would suggest no apparent influence of the change in rates between time-points for one phenomenon having an influence on the other. Figure 5.4 shows separate year by year correlations of infant mortality rate and total fertility rate. This shows lower correlations for infant mortality for successive years than for fertility. Year by year correlations of infant mortality fluctuate substantially, particularly during the 1980s; i.e. the distribution of infant mortality in successive years

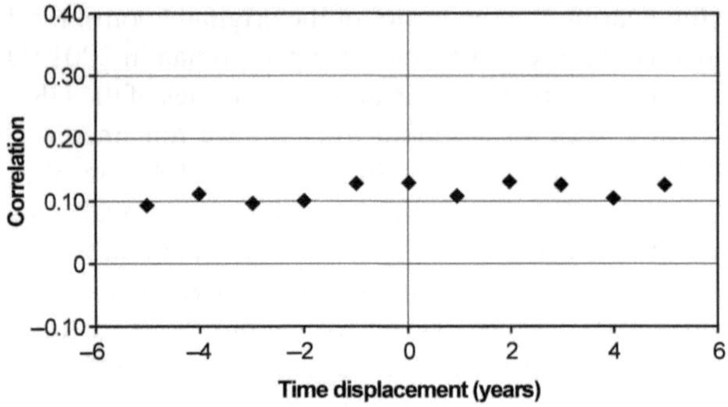

Fig. 5.3 Average correlations between changes in infant mortality and total fertility, 1981–2006 (Source: Author's calculations based on data supplied by ONS, GROS and NISRA)

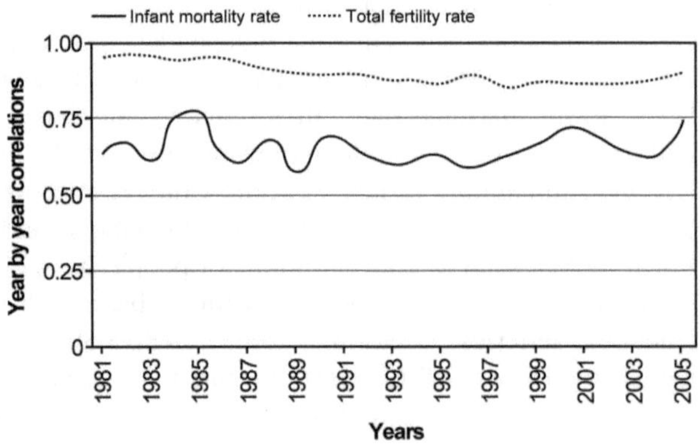

Fig. 5.4 Year by year correlations for infant mortality rates and for total fertility rates: 1981–2006 (Source: Author's calculations based on data supplied by ONS, GROS and NISRA)

is less consistent than the fertility rates in successive years. Fertility rates appear more consistent over time but the correlations become less strong as fertility rates reduce up to 2001 but increase when fertility rates recover up to 2006. When the first and last 5 years of this study period are compared, generally the relative position of areas is preserved to a fair extent over time since fertility rates for 1981–2005 and 2002–2006 correlate ($r = 0.52$, $p < 0.000$), although the relationship is less strong for infant mortality ($r = 0.32$, $p < 0.000$).

Since there is no maximum correlation at zero time displacement here, as in Brass and Kabir (1979), the suggestion that the causative factors may largely be the same for fertility as for child mortality seems unlikely. Nevertheless, there is a high degree of overlap for factors which are associated with differing levels of fertility and mortality, particularly the socio-economic variables (Richter & Adlakha, 1989; Woods, 1986b). Following a recent demonstration of the utility of modelling cross-sectional influences on fertility for areas over time (Boyle, Graham, & Feng, 2007), ordinary least squares (OLS) regression models are developed here for both 1991 and 2001 with infant mortality rates and total fertility rates as the dependent variables. In the models, explanatory variables are drawn from the 1991 and 2001 Censuses with the 1991 data adjusted because of boundary changes which occurred between the 1991 and 2001 (Norman, Simpson, & Sabater, 2008b) and so that variable definitions are comparable over time (Norman, 2010). Data are transformed to near normal distributions.

Table 5.1a shows a range of explanatory variables found to be significant within a regression model with total fertility rate in 1991 as the dependent variable. Higher deprivation (Townsend, 1987) and increased levels of persons of Pakistani or Bangladeshi ethnicity are associated with higher levels of fertility but an increase in the proportion of students and of persons of Chinese ethnicity are associated with lower levels of fertility. Compared with England, fertility is shown by dummy variables to be lower in Scotland, higher in Northern Ireland, but only marginally higher in Wales. Similarly, dummy variables on area type show that semi-rural and urban areas have successively lower fertility levels than rural areas. Table 5.2a shows that the same effects are still present in 2001 although the difference between semi-rural and rural areas is no longer significant.

To see whether infant mortality is influential on fertility levels, the 1991 infant mortality rate is added to the explanatory variables used in Table 5.1a. Whilst the effect is positive (i.e. a rise in infant mortality would be associated with a rise in fertility), the variable is not significant in the model (Table 5.1b). As there may be a time lag before the effect of infant mortality may apply, the model was reproduced but with 1996 total fertility rate as the dependent variable. The influence of infant mortality on fertility was positive but not significant in the model (Table 5.1c). Tables 5.2b, c have equivalent models for 2001 and for 2006 which show results consistent with those for 1991 and 1996.

Modelling infant mortality rate in 1991 as the outcome, higher deprivation and increased percentages of persons of low social class and persons of Pakistani or Bangladeshi ethnicity are all associated with increased levels of infant mortality (Table 5.3a). Table 5.4a shows the equivalent model for infant mortality in 2001. The same variables are found to be significant and with the direction and relative level of effect consistent with the 1991 model. Other factors including the separate input variables within the Townsend deprivation index (unemployment, overcrowded households, non-home ownership and lack of access to a car), qualifications, area type and others individually correlated with infant mortality in directions consistent with the literature but did not make a significant contribution to the regression models.

Various studies have found that increased levels of fertility can lead to higher levels of infant mortality (see for example, Dorsten, 1994; Talwalkar, 1981) rather than *vice versa*. To investigate this here, total fertility rate has been added to the set of explanatory variables reported in Table 5.3a. In Table 5.3b, it can be seen that increased total fertility rate has a small positive impact on infant mortality but this effect is not significant. Since there may be a time delay in fertility level having an effect, the dependent variable is changed to be the infant mortality rate in 1996. Table 5.3c shows that the model is a less good fit than before and that the influence of 1991 fertility is even less significant than with the demographic measures for the same year. In equivalent models for 2001 and for 2006 (Table 5.3b, c), fertility also has a non-significant effect.

Table 5.1 Modelled socio-economic influences on 1991 and 1996 total fertility rates

a Fertility 1991	R square	Adjusted R square	SE
Model summary	0.58	0.58	13.76
Dependent variable: 1991 Total fertility rate (log)			
	B	SE	Sig.
Constant	1.196	0.014	0.000
1991 Townsend deprivation index	0.004	0.001	0.000
% Students (log)	−0.072	0.008	0.000
% Pakistani or Bangladeshi ethnicity (log)	0.044	0.004	0.000
% Chinese ethnicity (log)	−0.099	0.012	0.000
Wales	0.024	0.009	0.007
Scotland	−0.064	0.008	0.000
Northern Ireland	0.119	0.012	0.000
Urban	−0.022	0.006	0.001
Semi-rural	−0.014	0.005	0.005
b Fertility 1991	**R square**	**Adjusted R square**	**SE**
Model summary	0.58	0.57	13.78
Dependent variable: 1991 Total fertility rate (log)			
	B	SE	Sig.
Constant	1.185	0.024	0.000
1991 Townsend deprivation index	0.004	0.001	0.000
% Students (log)	−0.073	0.008	0.000
% Pakistani or Bangladeshi ethnicity (log)	0.043	0.004	0.000
% Chinese ethnicity (log)	−0.097	0.012	0.000
Wales	0.024	0.009	0.006
Scotland	−0.064	0.008	0.000
Northern Ireland	0.120	0.012	0.000
Urban	−0.022	0.006	0.001
Semi-rural	−0.015	0.005	0.004
1991 Infant mortality rate	0.006	0.010	0.577
c Fertility 1996	**R square**	**Adjusted R square**	**SE**
Model summary	0.59	0.58	14.10
Dependent variable: 1996 Total fertility rate (log)			
	B	SE	Sig.
Constant	1.183	0.024	0.000
1991 Townsend deprivation index	0.004	0.001	0.000
% Students (log)	−0.096	0.008	0.000
% Pakistani or Bangladeshi ethnicity (log)	0.045	0.004	0.000
% Chinese ethnicity (log)	−0.053	0.013	0.000
Wales	0.036	0.009	0.000
Scotland	−0.077	0.008	0.000
Northern Ireland	0.114	0.013	0.000
Urban	−0.023	0.007	0.001
Semi-rural	−0.007	0.005	0.184
1991 Infant mortality rate	0.003	0.010	0.735

Notes: OLS regressions – weighted by persons.
Dummy variables for country are relative to England.
Dummy variables for area type are relative to Rural.

Table 5.2 Modelled socio-economic influences on 2001 and 2006 total fertility rates

a Fertility 2001	R square	Adjusted R square	SE
Model summary	0.57	0.56	15.59
Dependent variable 2001 Total fertility rate (log)			
	B	SE	Sig.
Constant	1.217	0.019	0.000
2001 Townsend deprivation index	0.004	0.001	0.000
% Students (log)	−0.113	0.010	0.000
% Pakistani or Bangladeshi ethnicity (log)	0.054	0.004	0.000
% Chinese ethnicity (log)	−0.050	0.011	0.000
Wales	0.030	0.010	0.002
Scotland	−0.045	0.008	0.000
Northern Ireland	0.090	0.013	0.000
Urban	−0.024	0.007	0.001
Semi-rural	−0.011	0.006	0.066
b Fertility 2001	**R square**	**Adjusted R square**	**SE**
Model summary	0.57	0.56	15.57
Dependent variable 2001 Total fertility rate (log)			
	B	SE	Sig.
Constant	1.200	0.022	0.000
2001 Townsend deprivation index	0.003	0.001	0.001
% Students (log)	−0.116	0.010	0.000
% Pakistani or Bangladeshi ethnicity (log)	0.052	0.004	0.000
% Chinese ethnicity (log)	−0.046	0.011	0.000
Wales	0.031	0.010	0.002
Scotland	−0.045	0.008	0.000
Northern Ireland	0.090	0.013	0.000
Urban	−0.025	0.007	0.001
Semi-rural	−0.011	0.006	0.055
2001 Infant mortality rate	0.013	0.008	0.114
c Fertility 2006	**R square**	**Adjusted R square**	**SE**
Model summary	0.51	0.50	19.82
Dependent variable 2006 Total fertility rate (log)			
	B	SE	Sig.
Constant	1.358	0.028	0.000
2001 Townsend deprivation index	0.003	0.001	0.009
% Students (log)	−0.158	0.012	0.000
% Pakistani or Bangladeshi ethnicity (log)	0.052	0.005	0.000
% Chinese ethnicity (log)	−0.040	0.015	0.006
Wales	0.029	0.013	0.022
Scotland	−0.053	0.011	0.000
Northern Ireland	0.075	0.017	0.000
Urban	−0.010	0.009	0.289
Semi-rural	−0.012	0.007	0.086
2001 Infant mortality rate	0.019	0.011	0.076

Notes: OLS regressions – weighted by persons.
Dummy variables for country are relative to England.
Dummy variables for area type are relative to Rural.

Table 5.3 Modelled socio-economic influences on 1991 and 1996 infant mortality rates

a Infant mortality 1991	R square	Adjusted R square	SE
Model summary	0.33	0.33	67.87
Dependent variable 1991 Infant mortality rate (log)			
	B	SE	Sig.
Constant	1.261	0.133	0.000
1991 Townsend deprivation index	0.014	0.003	0.000
% Low social class (log)	0.233	0.046	0.000
% Pakistani or Bangladeshi ethnicity (log)	0.118	0.016	0.000
b Infant mortality 1991	**R square**	**Adjusted R square**	**SE**
Model summary	0.33	0.33	67.94
Dependent variable 1991 Infant mortality rate (log)			
	B	SE	Sig.
Constant	1.210	0.177	0.000
1991 Townsend deprivation index	0.015	0.003	0.000
% Low social class (log)	0.222	0.052	0.000
% Pakistani or Bangladeshi ethnicity (log)	0.115	0.017	0.000
1991 Total fertility rate (log)	0.079	0.181	0.662
c Infant mortality 1996	**R square**	**Adjusted R square**	**SE**
Model summary	0.25	0.24	84.40
Dependent variable 1996 Infant mortality rate (log)			
	B	SE	Sig.
Constant	0.953	0.220	0.000
1991 Townsend deprivation index	0.015	0.004	0.001
% Low social class (log)	0.304	0.064	0.000
% Pakistani or Bangladeshi ethnicity (log)	0.091	0.021	0.000
1991 Total fertility rate (log)	−0.056	0.225	0.803

Note: OLS regressions – weighted by persons.

In the models reported here there is no apparent evidence of levels of infant mortality influencing levels of fertility (and vice versa). However, there is some overlap in the explanatory variables which affect levels of infant mortality and fertility; an area's level of deprivation as measured by the Townsend index and the percentage of persons of Pakistani and Bangladeshi ethnicity. The final section of this chapter investigates which areas have a similar experience of trends in infant mortality and fertility over time as well as similar socio-economic characteristics. This is achieved by inputting (as z scores) the annual time-series of infant mortality rates and total fertility rates for 1981–2006 and the deprivation levels and percentage of persons Pakistani and Bangladeshi ethnicity into a k-means classification (Vickers & Rees, 2006).

A seven 'cluster' solution has resulted in which areas with similar demographic trends 1981–2006 are grouped together. For clarity, the clusters are presented in

Table 5.4 Modelled socio-economic influences on 2001 and 2006 infant mortality rates

a Infant mortality 2001	R square	Adjusted R square	SE
Model summary Dependent variable 2001 Infant mortality rate (log)	0.36	0.36	90.01
	B	SE	Sig.
Constant	0.741	0.165	0.000
2001 Townsend deprivation index	0.024	0.005	0.000
% Low social class (log)	0.339	0.056	0.000
% Pakistani or Bangladeshi ethnicity (log)	0.113	0.019	0.000

b Infant mortality 2001	R square	Adjusted R square	SE
Model summary Dependent variable 2001 Infant mortality rate (log)	0.37	0.36	89.89
	B	SE	Sig.
Constant	0.939	0.215	0.000
2001 Townsend deprivation index	0.022	0.005	0.000
% Low social class (log)	0.370	0.060	0.000
% Pakistani or Bangladeshi ethnicity (log)	0.126	0.020	0.000
2001 Total fertility rate (log)	−0.308	0.215	0.151

c Infant mortality 2006	R square	Adjusted R square	SE
Model summary Dependent variable 2006 Infant mortality rate (log)	0.21	0.20	131.21
	B	SE	Sig.
Constant	−0.178	0.314	0.570
2001 Townsend deprivation index	0.007	0.007	0.338
% Low social class (log)	0.590	0.088	0.000
% Pakistani or Bangladeshi ethnicity (log)	0.103	0.030	0.001
2001 Total fertility rate (log)	0.078	0.313	0.803

Note: OLS regressions – weighted by persons.

Fig. 5.5 in two groups: clusters 1, 3, 4, 6, and 7 and clusters 2 and 5. In terms of total fertility rates over the period (Fig. 5.5a), five clusters of local authorities experience similar trends but at different levels. Cluster 1 has the lowest fertility of these and cluster 4 the highest fertility (above theoretical 'replacement level' in all years except 2000–2002). Clusters 2 and 5 are characterised by rapidly reducing fertility from relatively high rates in the early 1980s. For infant mortality, the relative position of the areas is maintained over time for clusters 4 and 7 which in Fig. 5.5b have the lowest and highest infant mortality rates. There is some changing of the relative positions of clusters 1, 3 and 6 but these areas comprise the mid-ground of rates. Clusters 2 and 5 start the time period with the highest rates in the UK. Both clusters experience a rapid decline in rates but more fluctuations are evident in cluster 5.

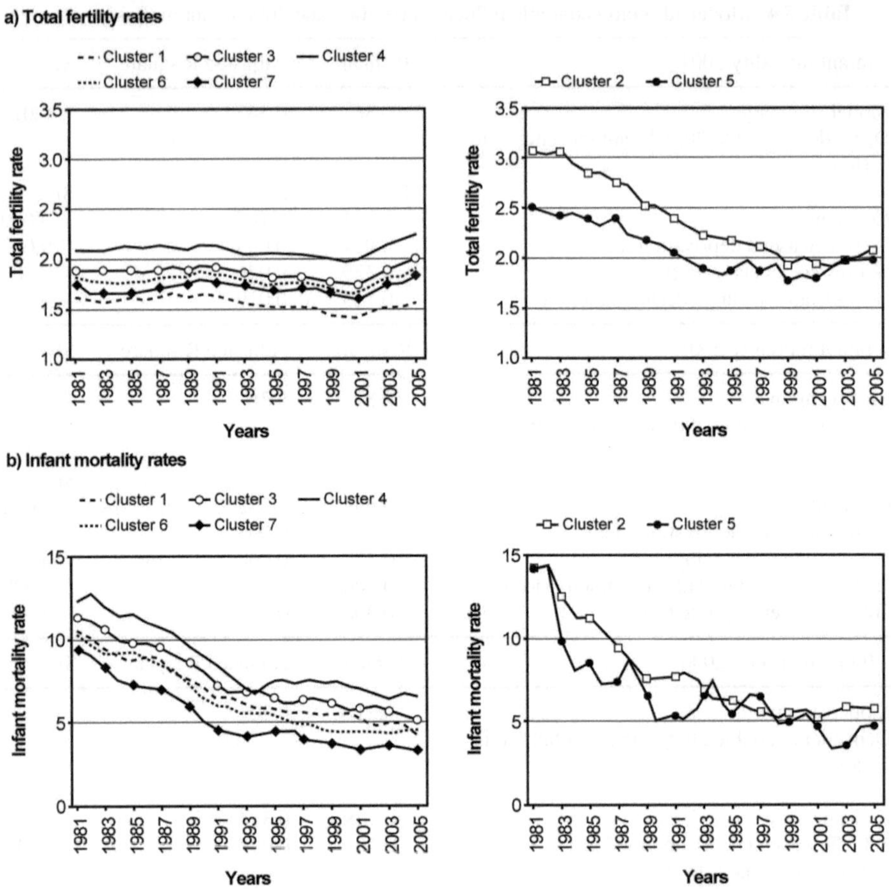

Fig. 5.5 A k-means classification of UK sub-national demographic trends, 1981–2006

In terms of the relationship between infant mortality and fertility trends for clusters, the lowest infant mortality cluster (7) does not have the lowest fertility but generally low infant mortality areas also have low fertility. Clusters 6 and 7 are the least deprived areas and these locations have low infant mortality. Whilst cluster 4 does experience an improvement in infant mortality rates, the rates remain relatively high and it is this cluster which retains relatively high fertility rates. Cluster 4 comprises areas with high percentages of persons of Pakistani or Bangladeshi ethnicity and which became relatively more deprived between 1991 and 2001. Clusters 2 and 5 are the most rapidly changing with rates of infant mortality and fertility declining rapidly through the period and being areas which experienced a marked reduction in the level of deprivation during 1991–2001.

The distributions of clusters in London and in Northern Ireland are illustrated in Fig. 5.6. There is a concentration of cluster 1 in inner London. This cluster is characterised by very low fertility but not a low level of infant mortality. This cluster is also found in major urban centres in Scotland's Central Belt and Leeds,

Fig. 5.6 Distribution of clusters in London and in Northern Ireland

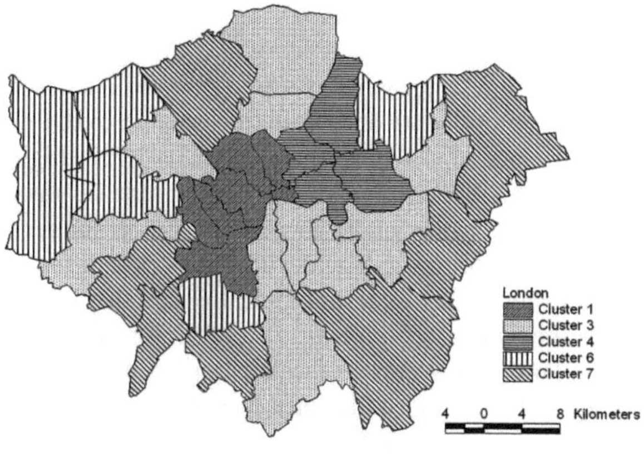

a London

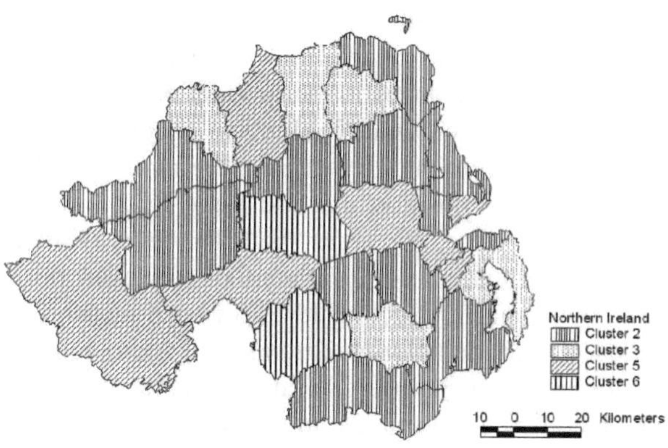

b Northern Ireland

for example. Away from inner London, cluster 3 has somewhat higher fertility and higher infant mortality and cluster 4 in North London is a concentration of relatively high fertility and infant mortality rates. This cluster is characterised by high percentages of people of Pakistani and Bangladeshi ethnicities and is grouped with locations such as Birmingham, Oldham, Rochdale and Bradford. Cluster 6 dominates West London and has a medium level of both fertility and infant mortality. Cluster 7 in East London has the second lowest fertility rates and the lowest infant mortality rates. Throughout the UK, areas grouped into cluster 7 include relatively rural and non-deprived areas.

Northern Ireland has areas classified into Clusters 3 and 6 as described above, but otherwise has clusters 2 and 5 which were shown in Fig. 5.5 to have very distinctive fertility and infant mortality trends with rapid declines in both measures during 1981–2006. Both clusters include accessible, semi-rural areas within commuting distance of urban areas, particularly Belfast. Cluster 5 also includes several more remote, rural areas. Clusters 2 and 5 are only found in Northern Ireland.

Conclusions

A range of demographic models can be used to highlight the contributions of different components on population change over time. The demographic transition model demonstrates how societies move from high to low mortality and fertility regimes. An underlying belief is that fertility decline follows mortality decline because people, perhaps subconsciously, no longer feel it necessary to have large numbers of children. A wide literature points towards reductions in fertility as a direct result of falls in infant mortality and this led Brass and Kabir (1979) to investigate the timing relationship of child mortality and fertility for subnational areas in England and Wales in the late nineteenth and early twentieth centuries. Since changes in child mortality rates were found to be contemporaneous with changes in fertility rates, Brass and Kabir concluded that the causative factors would largely be the same. Paradoxically, they noted geographical variations in child mortality trends but consistency in fertility trends.

In the late twentieth and early twenty-first centuries for sub-national areas across the UK, there is no correlation between changes in infant mortality rates and in fertility rates. Advances in the production of time-series socio-economic variables which are consistent in definition and geography over time do mean that influences on infant mortality and fertility can be modelled. There is some credence to Brass and Kabir's proposition that the same factors might affect both infant mortality and fertility since, for areas, both the level of deprivation and of Pakistani and Bangladeshi ethnicity are influential. However, further and different variables are also significant in affecting levels of each measure. Classifying areas by a combination of their demographic trends and socio-economic characteristics does reveal sets of areas which have similar experiences in their time-series of rates over time which could be informative in understanding trajectories of change. However, with no consistent geographical and timing patterns there is no clear evidence of a direct relationship between trends in infant mortality and fertility.

Acknowledgements Many thanks to Phil Rees for his unerring support and patient advice over the years. The research reported here is supported by ESRC Research Award RES-163-25-0032, 'What happens when international migrants settle? Ethnic group population trends and projections for local areas' and ESRC Research Award RES-189-25-0162. The Census data used were obtained via MIMAS' CASWEB and GIS boundary data via EDINA's UKBORDERS, facilities supported by ESRC and JISC. The Census, official Mid-Year Estimates and Vital Statistics for England and Wales, Scotland and Northern Ireland have been provided by ONS, GROS and NISRA and the digital boundary data by OSGB and OSNI. These data are Crown copyright and are reproduced with permission of the Office of Public Sector Information.

References

Botting, B., & Macfarlane, A. (1990). Geographic variation in infant mortality in relation to birthweight 1983–85. In M. Britton (Ed.), *Mortality and geography.* Series DS no. 9. London: HMSO.

Boyle, P. J. (2003). Population geography: Does geography matter in fertility research? *Progress in Human Geography, 27*(5), 615–626.

Boyle, P. J., Graham, E., & Feng, Z. (2007). *Contextualising demography: The significance of local clusters of fertility in Scotland* (Working Paper 2007-036). Rostock: Max Planck Institute for Demographic Research.

Brass, W., & Kabir, M. (1979). Regional variations in fertility and child mortality during the demographic transition in England and Wales. In J. Hobcraft & P. Rees (Eds.), *Regional demographic development* (pp. 71–88). London: Croom Helm.

Carstairs, V. (1981). Small area analysis and health service research. *Community Medicine, 3*, 131–139.

Chamberlain, J., & Gill, B. (2005). Fertility and mortality. In R. Chappell (Ed.), *Focus on people and migration* (pp. 71–89). Basingstoke: Palgrave Macmillan.

CICRED. (1975). *Seminar on infant mortality in relation to the level of fertility.* Paris: Committee for International Coordination of National Research in Demography.

Coleman, D. (2006). Immigration and ethnic change in low-fertility countries: A third demographic transition. *Population and Development Review, 32*(3), 401–446.

Dorsten, L. E. (1994). Direct and indirect effects on infant mortality in a high-fertility US population. *Population Research and Policy Review, 13*(1), 31–48.

Dunn, P. M. (2005). Sir George Newman MD (1870–1948) and the prevention of perinatal disease. *Archives of Disease in Childhood: Fetal and Neonatal Edition, 90*, F278–F280.

Dunnell, K. (2007). The changing demographic picture of the UK: National statistician's annual article on the population. *Population Trends, 130*, 9–21.

Fitzpatrick, J., & Cooper, N. (2001). Patterns and trends in stillbirths and infant mortality. In C. Griffiths & J. Fitzpatrick (Eds.), *Geographic variations in health* (pp. 161–190). London: The Stationery Office.

Freemantle, J., & Read, A. (2008). Preventing child deaths. Editorial. *British Medical Journal, 336*, 1083–1084.

García Coll, A., & Stillwell, J. (2000). *Internal migration and regional population dynamics in Europe: Spain case study* (Working Paper 00/8). Leeds: School of Geography, University of Leeds.

Garrett, E., Galley, C., Shelton, N., & Woods, R. (Eds.). (2006). *Infant mortality: A continuing social problem.* Aldershot: Ashgate.

Guildea, Z. E. S., Fone, D. L., Dunstan, F. D., & Cartlidge, P. H. T. (2005). Differences in risk of mortality under 1 year of age between rural and urban areas: An ecological study. *Public Health, 119*, 442–447.

Hobcraft, J., & Rees, P. (Eds.). (1979). *Regional demographic development.* London: Croom Helm.

Kmietowicz, Z. (2001). Health gap between rich and poor in London widens. *British Medical Journal, 323*, 828.

Kupiszewski, M., Durham, H., & Rees, P. (1996). *Internal migration and regional population dynamics in Europe: Polish case study.* Report prepared for the Council of Europe and the European Commission.

Lee, C. (1991). Regional inequalities in infant mortality in Britain, 1861–1971: Patterns and hypotheses. *Population Studies, 45*, 55–65.

Lee, R. (2003). The demographic transition: Three centuries of fundamental change. *Journal of Economic Perspectives, 17*(4), 167–190.

Lesthaeghe, R. J. (2005). *Second demographic transition.* Brussels: Vrije Universiteit. Online http://www.vub.ac.be/SOCO/Lesthaeghe.htm

Newman, G. (1906). *Infant mortality: A social problem.* London: Methuen & Co.

Norman, P. (2010). Identifying change over time in small area socio-economic deprivation. *Applied Spatial Analysis and Policy, 3*(2–3), 107–138.

Norman, P., Gregory, I., Dorling, D., & Baker, A. (2008a). Geographical trends in infant mortality: England and Wales, 1970–2006. *Health Statistics Quarterly, 40*, 18–29.

Norman, P., Simpson, L., & Sabater, A. (2008b). 'Estimating with confidence' and hindsight: New UK small area population estimates for 1991. *Population, Space and Place, 14*(5), 449–472.

Notestein, F. W. (1945). Population: The long view. In T. W. Shultz (Ed.), *Food for the world* (pp. 36–57). Chicago: Chicago University Press.
Office for Population Censuses and Surveys. (1978). *Occupational mortality. The Registrar general decennial supplement for England and Wales 1970–1972* (Series DS no. 1). London: HMSO.
Rees, P. (1986). Choices in the construction of regional population projections. In R. Woods & P. Rees (Eds.), *Population structures and models* (pp. 126–159). London: Allen & Unwin.
Rees, P. (1997). The second demographic transition: What does it mean for the future of Europe's population? *Environment and Planning A, 29*, 381–390.
Rees, P. H., & Convey, A. L. (1984). Spatial population accounting. In J. I. Clarke (Ed.), *Geography and population: Approaches and applications* (pp. 51–68). Oxford: Pergamon.
Rees, P. H., & Wilson, A. G. (1973). Accounts and models for spatial demographic analysis 1: Aggregate population. *Environment and Planning A, 5*, 61–90.
Rees, P. H., & Wilson, A. G. (1977). *Spatial population analysis*. London: Edward Arnold.
Rendall, M. S., & Smallwood, S. (2003). Higher qualifications, first-birth timing and further childbearing in England and Wales. *Population Trends, 111*, 18–26.
Richter, K., & Adlakha, A. (1989). The effect of infant and child mortality on subsequent fertility. *Journal of Population and Social Studies, 2*(1), 43–62.
Shelton, N. (2006). Conclusion: The social dimension to infant well-being. In E. Garrett, C. Galley, N. Shelton, & R. Woods (Eds.), *Infant mortality: A continuing social problem* (pp. 249–260). Ashgate: Aldershot.
Simpson, R. (2009). Delayed childbearing and childlessness. In J. Stillwell, E. Coast, & D. Kneale (Eds.), *Fertility, living arrangements, care and mobility understanding population trends and processes* (Vol. 1, pp. 23–40). Dordrecht: Springer.
Smith, S., & Ratcliffe, A. (2009). Women's education and childbearing: A growing divide. In J. Stillwell, E. Coast, & D. Kneale (Eds.), *Fertility, living arrangements, care and mobility understanding population trends and processes* (Vol. 1, pp. 41–58). Dordrecht: Springer.
Stone, R. (1971). *Demographic accounting and model building*. Paris: Organization for Economic Cooperation and Development.
Storeygard, A., Balk, D., Levy, M., & Deane, G. (2008). The global distribution of infant mortality: A subnational spatial view. *Population, Space and Place, 14*, 209–229.
Talwalkar, M. A. (1981). Association of infant mortality and high fertility: An empirical investigation. *IIPS Newsletter, 22*(1), 2–11.
Thompson, W. S. (1929). Population. *American Journal of Sociology, 34*, 959–975.
Townsend, P. (1987). Deprivation. *Journal of Social Policy, 16*, 125–146.
Tromans, N., Jefferies, J., & Natamba, E. (2009). Have women born outside the UK driven the rise in UK births since 2001? *Population Trends, 136*, 28–42.
Tromans, N., Natamba, E., Jefferies, J., & Norman, P. (2008). Have national trends in fertility between 1986 and 2006 occurred evenly across England and Wales? *Population Trends, 133*, 7–19.
van de Kaa, D. J. (1987). Europe's second demographic transition. *Population Bulletin, 42*(1), 1–59.
Vickers, D. W., & Rees, P. H. (2006). Introducing the national classification of census output areas. *Population Trends, 125*, 15–29.
Webb, J. W. (1963). The natural and migrational components of population changes in England and Wales, 1921–1931. *Economic Geography, 39*(2), 130–148.
Woodroffe, C., Glickman, M., Barker, M., & Power, C. (1993). *Children, teenagers and health: The key data.* Milton, Buckingham: Open University Press.
Woods, R. (1986a). The spatial dynamics of the demographic transition in the west. In R. Woods, & P. Rees (Eds.), *Population structures and models.* (pp. 21–44). London: Allen & Unwin.
Woods, R. (1986b). Spatial and temporal patterns. In R. Woods & P. Rees (Eds.), *Population structures and models* (pp. 5–20). London: Allen & Unwin.
Woods, R., & Rees, P. (Eds.). (1986). *Population structures and models.* London: Allen & Unwin.

Chapter 6
Monitoring Who Moves Where: Information Systems for Internal and International Migration

John Stillwell, Peter Boden, and Adam Dennett

Introduction

Migration is a concept that implies a change of usual residence involving individuals, couples, families or groups of people that we collectively refer to as migrants. The journeys that these individuals undertake between origin and destination locations may take place relatively quickly over very short distances – from one flat to another in the same apartment block – or involve extremely arduous experiences of travelling thousands of miles from one part of the world to another. The notion of a migrant therefore embraces a wide range of types of movement which may be voluntary or forced – as in the case of asylum seekers or displaced populations where the destination may be a temporary location rather than a permanent home. Although we frequently refer to international migrants to the UK as being either economic migrants or asylum seekers, this dichotomy conceals a plethora of migrant streams motivated to enter the UK for a wide range of reasons, not least to join with family members or to study at our institutions of higher education. Moreover, the distinction between shorter-distance residential mobility and longer-distance labour migration is sometimes used as a means to categorise domestic or internal migration in the UK when we know that many short-distance moves are motivated by marriage or divorce and long-distance streams include many retirement migrants, for example. Whilst the magnitude and composition of migration flows occurring within or across national borders of any country is likely to involve a complex 'bundle' of individuals with different motivations, it is also possible to identify streams for whom the drivers are the same, be they economic, political or social. In many cases, the common factor amongst the migrants that constitute a particular stream will be the stage of their life course; students, single workers, families with children, empty nesters and retirees are all familiar examples.

Given the wide variety of types of migration with different motivations taking place at different spatial scales, it is not surprising that there is no single source

J. Stillwell (✉)
School of Geography, University of Leeds, Leeds LS2 9JT, UK
e-mail: j.c.h.stillwell@leeds.ac.uk

J. Stillwell, M. Clarke (eds.), *Population Dynamics and Projection Methods*, Understanding Population Trends and Processes 4, DOI 10.1007/978-90-481-8930-4_6,

of data that provides comprehensive and reliable information about the volume, complexion and distribution of migration flows. The reality is that migration data are derived from many different sources, several of which were not created with the intention of providing migration statistics per se. Administrative records, such as those collected by the Department of Work and Pensions (DWP) or by the processing of National Health Service (NHS) patient registers, are now being used as indirect sources of migration data to supplement information collected directly from sample surveys such as the International Passenger Survey (IPS) or from the decennial Census of Population. We will outline briefly the main sources of census, survey and administrative data on international and internal migration before explaining and illustrating three migration information systems developed at the University of Leeds that serve as 'one-stop shops' for users to access different types of migration data.

The first of these systems is the *Web-based Interface to Census Interaction Data (WICID)*, a system developed and maintained by the Centre for Interaction Data Estimation and Research (CIDER), a data support unit funded by the ESRC under the Census Programme from 2001 to 2011 (Stillwell & Duke-Williams, 2003). *WICID* provides members of the UK academic community with online access to various internal and international data sets generated from previous censuses as well as certain administrative data. The second example is a spreadsheet-based database system known as the *New Migrant Databank (NMD)* that was initially recommended to the Greater London Authority (GLA) by Rees and Boden (2006) as a method of providing more accurate monitoring of international migration at local level. The *NMD* concept has been embraced by the Office of National Statistics (ONS) in the system they have developed for quick dissemination of the most recent international migration data. The *NMD* has also been implemented for use by the North West and Yorkshire Regional Migration Partnerships. The third system is known as the *European Demographic Databank* (*EDD*) and is currently under construction as part of an ESPON-funded project to analyse the 'Demographic and Migratory Flows affecting European Regions and Cities' (DEMIFER). The system, whilst not wholly dedicated to migration data, does feature migration as one of its key components, and contains migration statistics drawn from different European sources but also a considerable amount of estimated data given the lack of data available from existing sources at national and regional scales across the 31 states of Europe stipulated by ESPON. The main features of each of these systems will be explained and illustrated with examples of system interfaces or extracted data. However, we begin the chapter by considering why it is that robust migration statistics are so important.

Why Do We Need Better Migration Statistics?

We live in a world where securing reliable 'evidence' is an imperative. Planners and policy makers at all levels across many sectors require sound evidence upon which to base the decisions that they have to make regarding policy or service provision.

Monitoring internal and international migration has assumed higher priority because of the impact of both phenomena on the size and structure of local populations. Natural change, brought about by changes in births and deaths and reflecting trends in fertility and mortality or life expectancy are relatively well documented and reasonably predictable. Migration, on the other hand, whether internal or international, tends to be a more important component of population change and one whose trends are more difficult to predict than natural change components.

Every year, the ONS produces mid-year population estimates that require accurate information about migration flows. The 'official' or ONS sub-national population estimation methodology (NSCD, 2006) uses a number of alternative sources of immigration and emigration data to derive its estimates of Total International Migration (TIM) that are input to a cohort component model (Jefferies & Fulton, 2005), whereas the internal migration estimates are derived from NHS patient records (ONS, 2007a). In the case of the latter, ONS extracts annual data on migrants between local authority areas by comparing NHS patient registers held in the Patient Register Data System (PRDS) and then adjusting these estimates using patient re-registration data for health authority areas obtained from the NHS Central Register (NHSCR) for England and Wales. The sub-national population estimates are particularly important statistics not only because they are used for resource allocation across Government departments but also because they are used as denominators for key indicators such as gross domestic product (GDP) per capita. Sub-national population projections produced biennially incorporate trends in internal and international migration during recent historical periods (ONS, 2008) and these are also essential for planning local service provision. Good migration statistics are therefore required for financial and analytical reasons, as well as for monitoring the redistribution behaviour of the population. Yet, in the absence of a population registration system that records individual movements, migration statistics are difficult to measure accurately or to estimate with confidence. The problem is compounded by the issue of the differing duration of stay; ONS choose to use the United Nations (UN) definition of a long-term international migrant as someone who moves from their country of usual residence for a period of at least a year, whereas short-term migration is measured either as those who move for a period of 3–12 months for the purpose of work or study (the UN definition) or those moving for a period of between 1 and 12 months where the latter covers people who move for any reason, including tourist visits.

The dearth of migration statistics in Britain has been a longstanding problem but matters came to a head in the early years of the new millennium when several local authorities expressed concern over differences between the 2001 mid-year estimates rolled forward from 1991 and the 2001 Census population counts, leading to more general public concern over the population and migration statistics. The ONS identified weaknesses in their estimation of international migration and investigations began in 2003 with a National Statistics Quarterly Review of international migration statistics, whose report (ONS, 2003) heralded the start of a series of Parliamentary committee enquiries, reports and improvement programmes that are summarised in a recent report by the UK Statistics Authority (2009). These investigations were undertaken at a time when the European Union (EU) had expanded (in 2004) and the UK was experiencing almost unprecedented migrant inflows, largely from the eight Accession (A8) countries of central and eastern Europe. In 2005, the Statistics Commission produced recommendations that ONS and the Home Office should work together towards improving internal and international migration statistics, recognizing that 'some £100 billion a year is being distributed through formulae that are directly affected by migration estimates' (Statistics Commission, 2005, p. 506). Thereafter, following the publication of a report by an Inter-departmental Task Force on Migration (2006) which made a series of recommendations for the improvement of long-term and short-term migrants to the UK, ONS took the lead in establishing the Improving Migration and Population Statistics (IMPS) initiative in order to produce better passenger survey estimates of migration, better estimates of the geographic distribution of immigrants within the UK and better estimates of short-term migration as well as greater clarity in the production of population estimates, future surveys and the 2001 Census.

A House of Commons Treasury Committee reported in 2008 that the population estimates were 'not fit for purpose' and concluded that the International Passenger Survey (IPS) in its current state was unsuitable for estimating international migration properly, recommending that it should be replaced by a more comprehensive survey. A cross-government Migration Statistics Improvement Programme (MSIP) was established later in 2008 with five working groups created to address the recommendations of the earlier Task Force as follows:

1. *Entry and Exit* (led by the Home Office) to enhance the IPS and capture better information on immigration;
2. *Local population estimates* (led by Communities and Local Government, CLG) to improve migration estimates that underpin mid-year population estimates;
3. *Alternative sources* (led by the Department of Work and Pensions, DWP) to evaluate and develop possible new sources of migration statistics;
4. *Analysis and Indicators* (led by ONS) to develop an online information system containing up-to-date national and local migration indicators; and
5. *Migration reporting* (led by ONS) to coordinate the release of population and migration statistics across Government departments.

The MSIP programme has a 4-year time horizon (2008–2012) and runs alongside the CLG's own programme on 'Managing the Impacts of Migration' (CLG, 2008) which sets out to optimise the local benefits of migration and manage its transitional impacts by working across Government and with other stakeholders. On 1 April 2008, a House of Lords Select Committee on Economic Affairs report was published that reemphasised the necessity to 'improve radically the present entirely inadequate migration statistics' (p. 6) but also recognised that 'linking administrative databases held by different government departments can be difficult because of data protection and privacy issues as well as running the risk of losing data "in transit"' (p. 21).

Inadequacies in the provision of migration statistics are not confined to the UK. Bilsborrow, Hugo, Oberai, and Zlotnik (1997) review the wide range of conceptual and analytical issues related to the measurement of stock and flows of international migrants worldwide, together with problems concerning international comparability of migration data. They discuss the potentials and weaknesses of existing data collection systems – including censuses, population registers, border statistics, and residence and work permit systems – and provide guidance on how to develop and disseminate statistics on international migration. UN recommendations on international migration statistics (United Nations, 1998) provide a useful target for improving the collection, reliability and comparability of such statistics and Poulain and Perrin (2003) have explored the current state of European migration statistics and the prospects for greater coordination. They emphasise that collection of migration data is often a by-product of administrative data collection systems and that across Europe, countries often do not have the same political, economic or social interest in collecting data on immigration and emigration. EU citizens can now live in other European countries without asking for a residence permit to stay and migration within Europe is sometimes considered as irrelevant by administrations in charge of migration management whose main attention is directed towards immigrants from outside the EU, asylum seekers, and illegal migrants. Counts of the migration flows between EU states vary according to how the migrants are measured by the countries concerned and the problem of harmonising definitions and concepts of migration is as important as improving the data collection systems. Eurostat, for example, in working towards a more consistent definition of short-term and long-term migrants, is looking towards the UN recommendations mentioned above.

Whilst the problem of international comparability of migrant counts between countries has been dealt with by Eurostat through bilateral agreements between certain countries, an alternative approach has been to develop a mathematical model to harmonise and estimate incomplete or missing data in the inter-state migration matrix (Raymer, 2008). Raymer and Abel (2008) document the multiplicative component model and iterative fitting methods developed in their MIMOSA project which made use of reliable data on migration between countries to estimate flows where the data were less reliable or non-existent.

Sources of Internal and International Migration Data in Britain

There is a key conceptual difference between 'migrants/movers' – the individuals involved in moving from one residential location to another – and 'migrations/moves' – the events of transfer between origins to destinations. Rees (1977) identified migration data from the census as counts of migrants or 'transitions' since the individuals involved were in existence at both the beginning and the end of the measurement period; i.e. they are people who survive the period and who are at a different usual address on Census date than that 12 months previously, regardless of whether they have made single or multiple moves. Migrants therefore contrast with the count of moves or transfers taking place during a time period; i.e. administrative

Table 6.1 Internal and international migration data sources

Internal migration	International migration
Censuses (various products)	
Main tables	Main tables
Special migration statistics (SMS)	Special migration statistics (SMS)
SARs (including SAM and CAMS)	SARs (including SAM and CAMS)
Longitudinal studies	Longitudinal studies
Commissioned tables	Commissioned tables
School census (now triennial)	School census
Surveys	
General household survey (GHS)	International passenger survey (IPS)
Labour force survey (LFS)	Labour force survey (LFS)
Annual population survey (APS)	
Acxiom research opinion polls	
Administrative records	
NHSCR	
Patient registration data system (PRDS)	GP registrations (Flag 4 data)
Higher education statistics agency (HESA)	Higher education statistics agency (HESA)
Electoral roll/register	Electoral roll/register
	National insurance numbers (NINo)
	Works and pensions longitudinal study (WPLS)
	Worker registration scheme (WRS)
	Seasonal agricultural workers (SAW)
	Immigration controls
	Work permits
	Points-based system
	Landing cards

records may measure movement events rather than individual migrants and therefore capture multiple and return moves. The difference between transition and movement concepts underpins the distinction between data on internal and international migration in Britain that are available from the Census and those data available from administrative sources such as the National Health Service Central Register (NHSCR), and survey sources such as the International Passenger Survey (IPS). A recent audit of origin-destination migration data is reported in Dennett, Duke-Williams, and Stillwell (2007) which describes the variety of census products, administrative databases and national surveys from which migration flow data can be extracted. Summary tables of data sources together with detailed synopses of each dataset are reported in Rees, Stillwell, Boden, and Dennett (2009). Table 6.1 therefore provides a single overall summary of the various sources of data that are available. The suite of *Census* macro and micro data products only provide information about immigration and not about emigration. The *School Census* which has evolved from the Pupil Level Annual School Census (PLASC) and is now collected three times a year, provides geo-references of pupil home locations which can be compared longitudinally to derive counts on internal and immigration of those of school age.

Migration data can be extracted from the results of various surveys. The *Labour Force Survey* (*LFS*) is a quarterly sample survey of households living at private addresses in the UK and asks the question: 'Where were you living 1 year ago?', providing a count of the flow of internal and international migrants within a single year. The *General Household Survey* (*GHS*) is a continuous household survey of 20,000 individuals from which data on inter-regional migration are available. The relatively new *Annual Population Survey* (*APS*) combines the results from the LFS waves 1 and 5 with the English, Welsh and Scottish Local Labour Force Survey (LLFS). The APS sample size is 122,000 households (or 375,000 respondents) compared with the quarterly LFS sample of 57,000 households (or 120,000 respondents). Estimates of non-UK born population and non-British citizens extracted from the APS provide an indication of migration stocks. The *International Passenger Survey* (*IPS*) is the only instrument in the UK which asks questions about immigration and emigration. It samples approximately 250,000 passengers each year, of which about 1% are migrants whose stated intention is to stay in or leave the UK for more than 12 months. This is equivalent to approximately 3,000 migrant respondents, 70% of whom are immigrants and 30% are emigrants. From 2007, the number of interviews with departing migrants has been boosted to a comparable level to those on entry. In contrast to these public sector data sets, *Acxiom* is an international private company that carries out an extensive opinion poll survey twice a year. As well as asking questions about consumer behaviour, the Acxiom questionnaire for certain years also contains questions about past migration as well as migration intentions.

Migration data can be sourced from a range of administrative registers including the *NHSCR* and the *PRDS* mentioned previously. An *Electoral Register* is maintained by each local authority and is an annual list of all individuals resident in an area who are eligible to vote. Migration data can be generated by comparing the names on the electoral roll from 1 year to another. The *Higher Education Statistics Agency* (*HESA*) collects data on all students entering, studying in and then leaving the UK Higher Education (HE) system. There are two main sources of HESA data: the 'Student' data set records parental domicile (unit postcode) and institution of study for all UK domiciled students for each year of study and the 'Destinations' data set contains all individuals previously recorded in the 'student' dataset who leave the UK HE system following completion of study. HESA administrative systems do not capture the residential address of international students, only the location of the institution of study. *Flag 4 GP registration* records are provided to ONS from the PRDS data provided by NHS Connecting for Health (NHSCfH). Flag 4s are codes within the PRDS system which indicate that someone who has registered with a GP in England and Wales was previously living overseas. A Flag 4 may be generated when an individual registers with an NHS GP if an individual was born outside the UK and enters England and Wales for the first time and registers with a NHS GP. An individual's registration will also generate a Flag 4 if the previous address of an individual is reported as outside the United Kingdom, and time spent outside the UK is more than 3 months. *National Insurance Number* (*NINo*) *statistics* managed by the Department of Works and Pensions (DWP) are allocations to overseas nationals entering the UK who are over 16 years of age, planning to work or claim benefits but asylum seekers are not eligible until their cases have been approved. The *Works and Pensions Longitudinal Study* (*WPLS*) provides data on the benefit claims and pensions of migrants whereas the *Worker Registration Scheme* (*WRS*) provides a count of the number of registrations of A8 citizens when they obtain a job in the UK and the *Seasonal Agricultural Workers* (*SAW*) data relate to the scheme which allows seasonal agricultural workers from other countries to come and work in the UK for a limited period of time, particularly during the planting and harvesting seasons.

In addition to these data sources and other sources of administrative data (such as details of work permits and immigration controls held by the Home Office, landing cards, or data from the points-based system maintained by the UK Border Agency), there are also composite data sets such as the Total International Migration (TIM) series mentioned earlier that are derived by ONS from various sources and are key inputs to the mid-year population estimates. These are discussed in a later section. The summary information in Table 6.1 outlines the variety of different sources of migration data that exist. However, even with the sources identified, there are frequently additional problems of locating, accessing and downloading the particular sets containing migration data for use in research and analysis. Moreover, origin-destination flows are notoriously difficult to handle because each count involves two locations and flow matrices can be very large but sparsely populated. In the remainder of the chapter, we seek to outline three migration information systems that have been designed and developed to draw migration data from different sources together and provide user-friendly interface and query facilities to allow users either to extract the data for processing or to display and monitor trends over time based on raw counts or derived indicators. All the examples that we have selected relate to the Government Office Region (GOR) of Yorkshire and the Humber or its component parts at ward, district or NUTS 2 region level.

Migration Information Systems

Web-Based Interface to Census Interaction Data (WICID)

The *WICID* system has been designed to enable members of the UK academic community to have access to the Special Migration Statistics (SMS) from the 1991 and 2001 Censuses and to encourage the use of the ESRC's investment in interaction data. In constructing *WICID* it was fundamental to provide a user-friendly interface, facilitating query building, quick data extraction and simple downloading of files containing interaction flows in common formats for analysis in other software packages. It was also necessary to provide a system that was open, in the sense that new data sets could be added as they became available without having to undertake any major system redesign. In its latest version, *WICID* contains annual data time series of inter-district migration flows in England and Wales, estimated by ONS from comparing annual NHS patient registers and constraining the flows to totals from the NHSCR. The creation of a generic framework of metadata files embracing

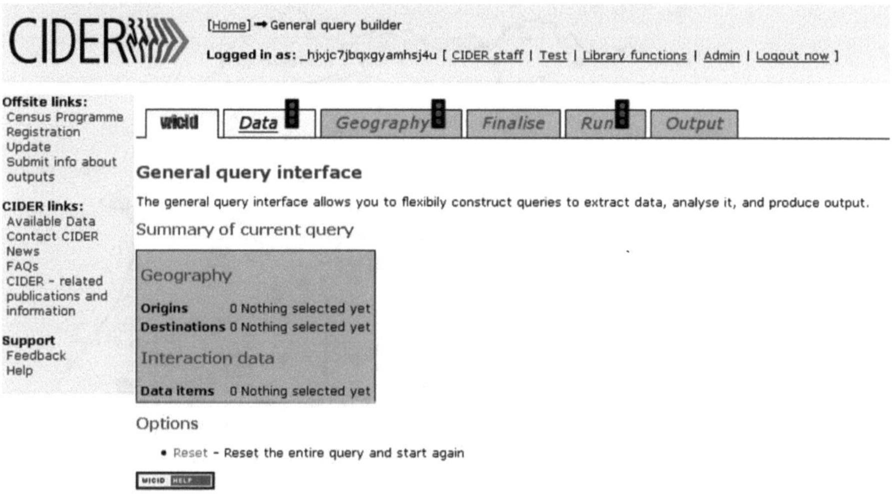

Fig. 6.1 The general query interface in WICID with nothing selected (Source: Screenshot from WICID system)

all the information components within the system has been particularly important in this context and detailed explanation of the metadata framework is found in Stillwell, Duke-Williams, Feng, and Boyle (2005). The system architecture has been described by Stillwell and Duke-Williams (2003), issues relating to the 2001 Census interaction data are discussed in Stillwell and Duke-Williams (2006) and details of the interface together with examples of analyses based on data extracted from *WICID* are found in Stillwell, Duke-Williams, and Dennett (2010). Figure 6.1 illustrates the general query interface in *WICID* which uses a traffic light metaphor to ensure that the variable (data item) selections and the geography (origin and destination) settings have been completed satisfactorily before the query is allowed to run and output is downloaded.

In technical terms, one of the critical challenges in developing *WICID* was to allow users the flexibility of selecting origin and destination zones from different geographical scales. In practice, this means that the system is able to respond to the following type of query: extract all the migration flows by each ethnic group from districts in Yorkshire and Humber to other GORs in the UK; or extract all the migration flows into wards in Leeds from other wards in Leeds, from other surrounding districts in Yorkshire and Humber, from other GOR in the UK, and from abroad. We show two examples of data that have been extracted from *WICID* and used for analysis. The first illustrates the spatial variation that exists in net migration at the ward level across Yorkshire and the Humber. Just over 460,000 individuals moved within the region in the 12 months before the 2001 Census, 9.4% of the nearly 4.9 million household residents in 2001. Figure 6.2 shows two representations of net migration between wards within the region that were extracted from WICID.

Fig. 6.2 Net migration balances and rates by ward in Yorkshire and Humber, 2000–2001 (Source: Migration data from 2001 SMS Table 201 via *WICID*; population data for rate calculations from Standard Table via *CASWEB*)

The map on the left-hand side illustrates the urban orientation of the pattern of net migration losses and gains. The majority of the wards in the districts that constitute the West and South Yorkshire conurbations, together with Hull, York, Harrogate and Grimsby in North East Lincolnshire, show gains or losses of larger magnitude than those experienced by the more rural wards of North Yorkshire and the East Riding. On the other hand, the map on the right-hand side of Fig. 6.2 indicates that rates of migration tend to be higher in wards of North Yorkshire as well as in certain inner city wards. Aggregate net migration maps of this type demonstrate the complexity of the spatial patterns of migration across one region at this spatial scale. Interesting variations appear at the ward level that are obscured at the district scale where net migration losses occur from those districts classified by Vickers, Rees, and Birkin (2003) as 'regional centres' (e.g. Leeds and Sheffield) and the largest net gains are found in more rural districts and 'prospering smaller towns' (e.g. East Riding), in conformity with the national pattern of counterurbanisation for migration of all ages. The patterns vary regionally and nationally, of course, by demographic characteristics such as age (Dennett and Stillwell, 2010) and ethnic group (Stillwell & Hussain, 2010).

The Census therefore provides reliable and comprehensive data for small areas across the country on internal migration and immigration but its periodic nature is very limiting as far as monitoring migration over time is concerned. In order to establish how migration volumes and patterns change over time, ONS has traditionally obtained data on patient re-registrations from the NHSCR to produce tables of moves between health-related areas: Family Practitioner Committee (FPC) areas, Family Health Service Authority (FHSA) areas and Heath Authority (HA) areas most recently (see Stillwell, Rees, & Boden, 1992; Dennett, Duke-Williams, & Stillwell, 2007, for a review of NHSCR data). These can be aggregated to flows at the regional scale. However, the creation by ONS of a PRDS that allows comparison of records in 1 year with those of the previous year by linking on NHS number enables identification of each person who changes their postcode. Since individuals must be present in data from both years, this facilitates matrices of 'transition' data to be prepared for local authority areas as well as HA areas, with NHSCR data used to adjust the estimates to take account of flows missing from the patient register data. One of the products in the public domain is a table (ONS table 2a) of aggregate movements between unitary and local authorities in England and Wales, with no within area moves included. Data are rounded and there is no age-sex breakdown. However, flows between local authority areas are available from ONS disaggregated by eight broad age groups. Figure 6.3 illustrates how the volumes of age-specific net migration change over an 8 year period, mid-1998/mid-1999 to mid-2005/mid-2006 for the GOR of Yorkshire and Humber as a whole. The set of all age net migration balances for each of the years is shown on the right-hand side of the graph indicating how early losses have been transformed into gains in the first half of the 2000s, peaking in mid-2003/mid-2004. The sets of histograms to the left show the net migration balances in each year decomposed by broad age group. These data can be interpreted in terms the stages in the life course associated with particular ages. Thus, we note that the region gains migrants in the family age ranges (0–15

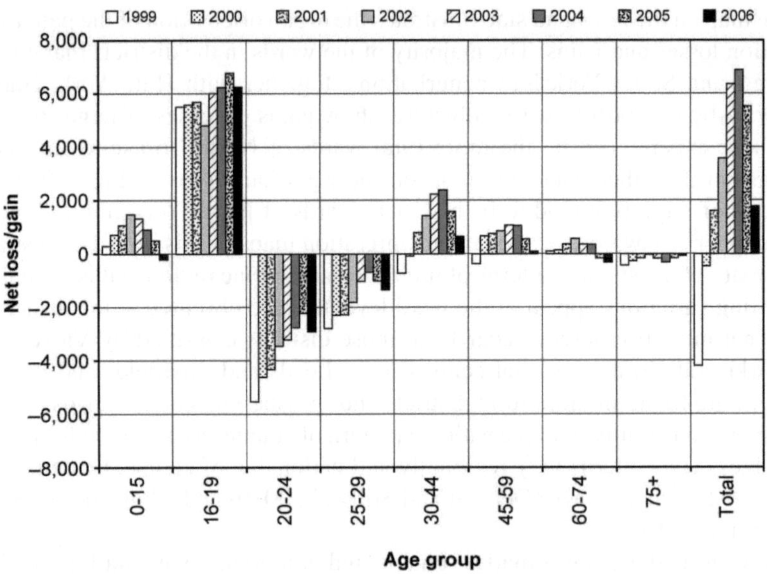

Fig. 6.3 Yorkshire and the Humber net migration losses and gains by age, 1999–2000 (Source: Patient register/NHSCR estimates provided by ONS and obtained from WICID)

and 33–44), in the student age group (16–19) and in the older working and retirement groups (44–74) but loses migrants in their twenties and in very old age (75+). The size of the 16–19 age group suggests the relative importance of student net in-migration as a component of population growth whereas the region is continuing to lose young labour force migrants in their twenties which will include the outflow of graduates. Although the net balance remains negative, the trend is encouraging from an economic planning perspective because the region appears to be experiencing a declining net out-migration and is therefore retaining more of its young adult workforce.

The time series PRDS/NHSCR data show us that London remains the preferred location for young out-migrants. This is no surprise; the capital has been attracting young migrants from all regions across the UK with its many tertiary and quaternary sector job and socio-cultural opportunities. However, the strong pull of the capital appears to have weakened between 2002 and 2005, when there was a sharp drop in the net out-migration balance from Yorkshire and the Humber, although the net rate of loss had increased again by 2006.

New Migrant Databank (NMD)

In this section, the focus turns from internal to international migration data. As Table 6.1 has indicated, there is no single instrument for the data collection and monitoring of international migration. The Census provides comprehensive spatial information about migrants who arrive in the previous 12 months from overseas but, as with internal migration, the data are cross-sectional and soon become out of date. Surveys are often rich sources of data but are typically not statistically robust for local-area analysis since they are based on relatively small samples and do not adequately capture all the populations that move into or out of the country. Administrative sources can provide excellent geographical detail of immigrants but typically do not have the data richness that surveys provide. Whilst reasonable time series data sets on immigration to the UK are available from a number of survey and administrative sources, there is a paucity of sources of data on emigration. Consequently, we know a good deal more about those people who arrive in the UK from abroad than we do about those who leave our shores to live in foreign places.

Total International Migration (TIM) statistics produced by ONS have been the definitive source of estimates on international migration to and from the UK in the past. The population estimation methodology uses a number of alternative sources of data to derive these estimates (NSCD, 2006), the principal source being the IPS, which samples around 300,000 passengers travelling through 34 air and sea ports (ONS, 2009) in Britain every year – of which around 1% can be considered migrants (NSCD, 2006; Jefferies & Fulton, 2005). Adjustments are made to the IPS data to account for 'visitors switchers', i.e. visitors who stay for longer than 12 months and become migrants, and 'migrant switchers', migrants who do not stay as long as expected and become visitors. The IPS is supplemented with data from the Irish Central Statistical Office and NHSCR which give more accurate information on migrants travelling to and from the Republic of Ireland. Data on asylum seekers and their dependents, not recorded by the IPS are supplied by the Home Office.

The concept of the *New Migrant Databank (NMD)* was originally recommended to the Greater London Authority (GLA) as a solution to its requirement for a system to measure and monitor international migration at a local level more accurately (Rees & Boden, 2006). The GLA has now produced its own descriptive analysis of international migration comparing TIM estimates with NINo registrations and GP registration statistics (Hollis, 2008). The development of the *NMD* has been taken forward by a team at the University of Leeds to produce a unique repository of migration statistics for a hierarchy of geographical areas in the UK, from national to local authority level (Boden & Rees, 2010a, 2010b). The driver for the creation of the *NMD* was the need to provide a single source of migration statistics for each local authority but also to facilitate the development of alternative migration estimation methods, specifically for international migration. The system has been adapted to support the work of the Regional Migration Partnerships in Yorkshire and Humber (Tyler & Boden, 2010) and the North West and the example of the system presented in Fig. 6.4 relates to one local authority selected by the user, the Metropolitan District of Leeds.

The *NMD* draws together statistics from a number of alternative sources and provides a consistent view of each local authority, benchmarked against regional and national totals. Since the data are captured at an aggregate geography, the issues associated with data protection and disclosure are minimal. The *NMD* juxtaposes datasets so as to identify the differences in the magnitude and trends observed from

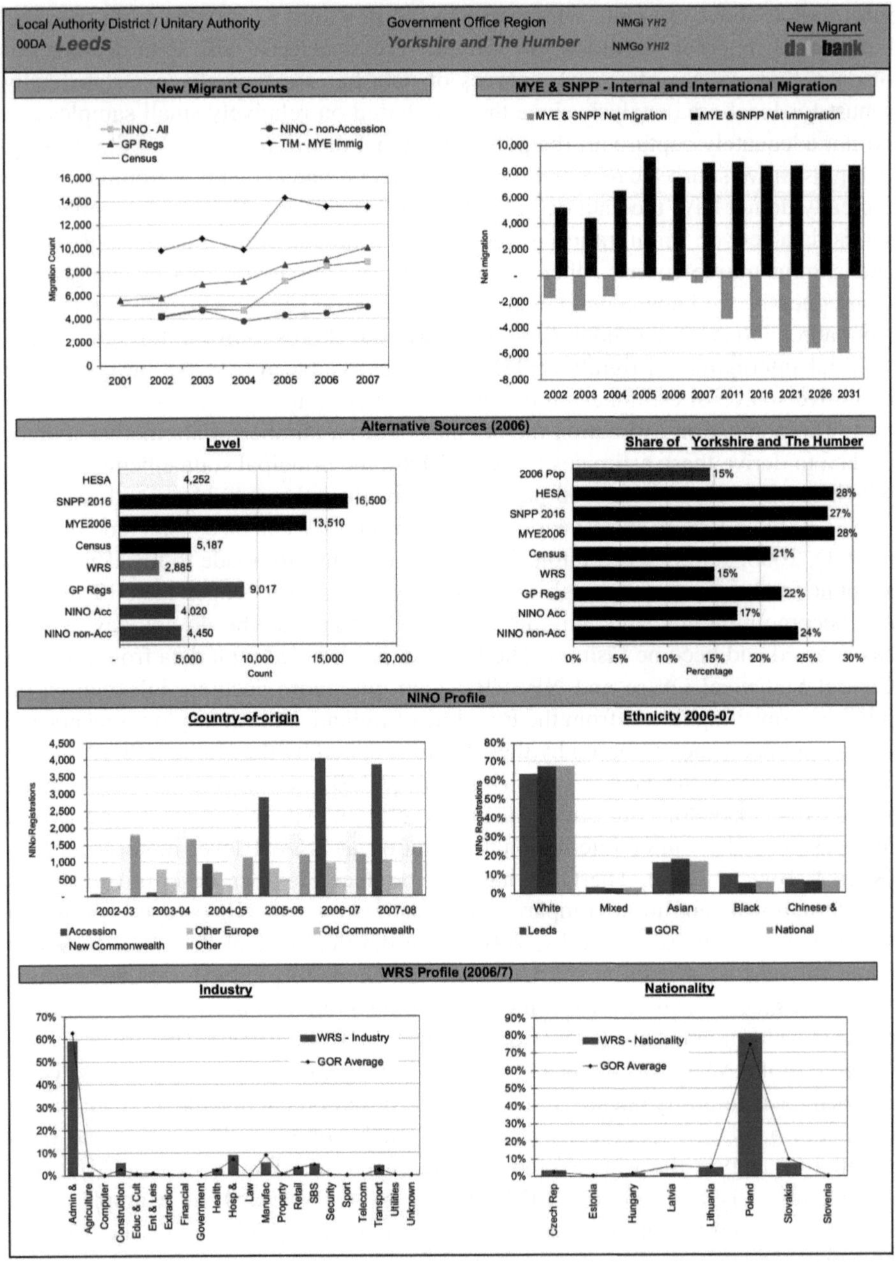

Fig. 6.4 *NMD* area profile: an example illustration for Leeds. All data are Crown copyright (Sources: 100% data extract from the National Insurance Recording System (NIRS): 2006 mid-year estimates (ONS, 2007a); 2006-based SNPP, current data (ONS, 2008); GP registration statistics provided by ONS; Workers registration scheme (Home Office, 2008))

each series. The initial summary chart labelled New Migrant Counts in Fig. 6.4 illustrates immigration trends over time and the differences that exist between data counts from four alternatives sources: the 2001 Census, TIM statistics, GP registrations of foreign nationals and NINo registrations of foreign workers. NINo data for non-Accession migrants are shown as well as the total number of registrations. The 2001 Census benchmark is represented as a constant across the time series. The data for Leeds indicate a steady increase in GP registrations from 5,500 in 2001 (which is very consistent with the recorded Census total for immigration) to over 10,000 in 2007. Throughout the period, however, TIM estimates for Leeds are consistently higher than GP registration totals. NINo registration totals for non-Accession workers remain relatively stable, whereas registrations by Accession migrants almost double from 2004 to 2007.

The second chart in Fig. 6.4 illustrates the net flow picture for both internal and international migration drawn from the migration assumptions that underpin the mid-year population estimates (MYE) produced by ONS for 2002–2006 and for the sub-national population projections (SNPP) for 2007–2031. Whilst the international net migration values are derived from the TIM statistics, the values for internal net migration are based on evidence from GP registrations between local authority areas in the UK. The profile for Leeds suggests a net gain due to international migration to 2007, continuing throughout the 25-year population projection horizon. In contrast, there is a small net loss due to internal migration which is expected to increase to 5,000–6,000 per year to 2031.

The two charts on the second row are designed to give a snapshot of the migrant counts for any single year that are produced from alternative sources and to illustrate whether the local authority area has a consistent share of its regional total. The first chart shows the extent of variation in the counts of migrants into Leeds in 2006 whereas the second chart shows the flows into Leeds as a proportion of the flows into Yorkshire and Humber GOR. In this instance, the charts display counts of student inflows from the HESA, TIM and SNPP estimates and the 2001 Census count, together with WRS, GP and NINo registrations for both Accession and non-Accession migrants. The key feature of these charts for Leeds is the difference between the TIM statistics, which are used in the MYE and SNPP, and the administrative statistics. Leeds' share of administrative registrations is much lower than the share of TIM migrant numbers that are allocated to Leeds.

Two charts based on NINo data are presented in the third row. The first gives an indication of the changing profile of migrant workers registering for a NINO since 2002–2003. This emphasises the sharp rise in Accession migrant registrations post-2004, coupled with relative stability in the level of registration for other country-of-origin groups. The second chart gives an illustration of the ethnic profile of NINo registrations for 2006–2007, derived by combining NINo country-of-origin data with corresponding data from the 2001 Census on international migration by ethnic group and country of origin. 63% of NINo registrations in 2006–2007 were classified as 'white' ethnic, which compares to a Census total of 60%, reflecting the rise in Accession and 'other Europe' migrant registrations to this local authority.

Finally, the bottom row of the profile contains WRS records of Accession migrants who work within the local area (though they may live elsewhere) by industry and by nationality. The vast majority of new Accession migrants to Leeds were employed in administrative and manual jobs. Approximately 80% were from Poland.

The development of the *NMD* has had two particularly important repercussions. First, it provides an illustration of how alternative measures of international migration can be validated and in doing so, how they can become an integral part of the methodology for mid-year population estimation. Second, it provides an example of how to construct an interactive interface to the alternative data sets, allowing users to select and display the data for a local authority of their choice. The first of these issues has been documented in the recommendations by Rees et al., (2009) for the UK Statistics Authority (2009) and in Boden and Rees (2010a). Figure 6.5 appears in the Appendix of Rees et al., (2009) and is a graph showing the total immigration to England over 2001–2007 as measured using statistics from the sources mentioned above. Although the TIM and GP statistics measure migration in different ways, the general trend of both these data sets is reasonably consistent and suggests that GP registrations might be a useful comparative measure of the level of long-term immigration. However, when the TIM inflow and GP registrations data are compared for GORs, there is much less consistency, particularly in the West Midlands, where the TIM estimates were 34% lower than the GP registrations in 2005–2007 and in Yorkshire and Humber and the South West, where they were 16% higher (Boden & Rees, 2010a).

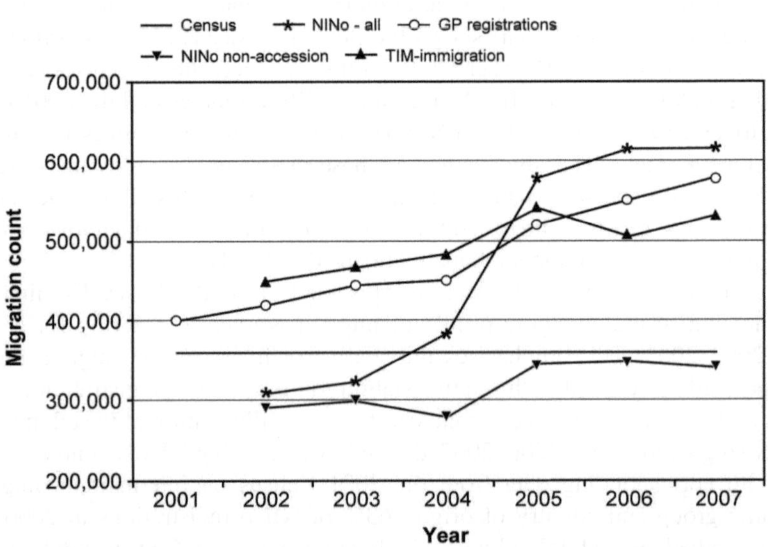

Fig. 6.5 Immigration trends from alternative data sources, England, 2001–2007 (Source: Rees et al., 2009, Appendix 3, p. 119)

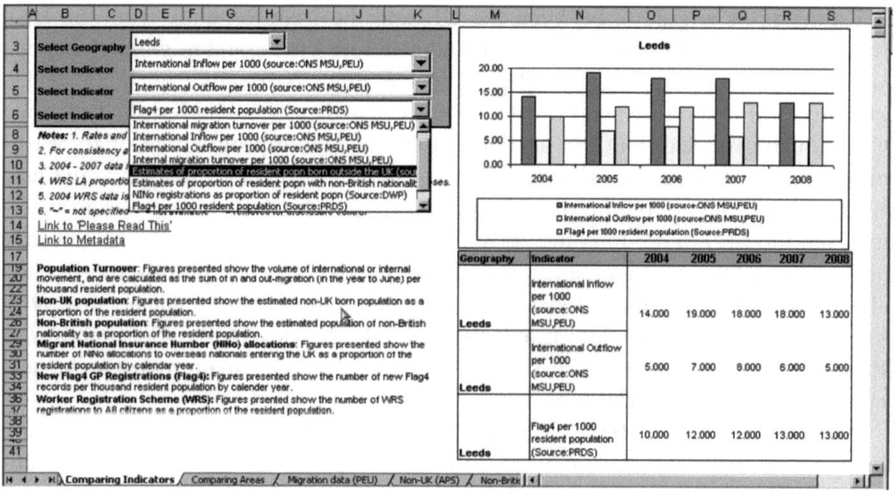

Fig. 6.6 Local migration indicator system showing selected indicators for Leeds (Source: http://www.statistics.gov.uk/statbase/product.asp?vlnk=15239)

Inconsistencies are similarly observed at the local authority level as shown in Fig. 6.6 which is an example of the Excel-based system built by ONS for local migration indicators whose development has been guided by the Analysis and Indicators Working Group of the MSIP. The system is available online from the National Statistics web site at http://www.statistics.gov.uk/statbase/product.asp?vlnk=15239. Although different in design from the *NMD*, the ONS system demonstrates the main principal underpinning the *NMD* of assembling and displaying time series data from alternative sources within the same interface. The local migration indicators system allows users to compare up to three indicators for a selected geographical area or to compare up to three geographical areas for one selected indicator (not shown in Fig. 6.6). The screenshot captures the list of indicators available from the dropdown menu that include turnover, inflow and outflow rates per 1,000 residents, estimates of the proportion of the resident population born outside the UK and the proportion with non-British nationality, NINo registrations, Flag 4 registrations and WRS registrations. The counts of the non-UK born and the non-British nationality populations come from the Annual Population Survey (APS) conducted by ONS. Users have access to the spreadsheets containing all the indicators for all the available geographies for their own analyses and metadata on all the indicators are available.

In the example shown in Fig. 6.6, TIM estimates of inflow (and outflow) rates are compared with Flag 4 rates per 1,000 residents for Leeds and show considerable variation. The TIM estimates are higher than the GP registrations in each year apart from 2008. These differences return us to the initial question of how the TIM estimates of immigration have been derived for sub-national areas. As mentioned earlier, TIM estimates combine data from the IPS on inflows and outflows with additional statistics on asylum seekers and their dependants from the Home Office, on estimates of migration between the UK and the Irish Republic from the Irish Central Statistical Office (ICSO) and are adjusted for visitor and migrant switchers. Analyses based on data from the *NMD* (Boden & Rees, 2010a) indicate that the TIM estimates of immigration are prone to error because they do not utilise the full range of administrative data that are available. ONS' ongoing programme of improvement of its international migration statistics has included some changes to the way sub-national estimates of long-term migration are produced. By incorporating statistics from the LFS rather than the 2001 Census to calibrate IPS migration flows for the London GOR and by creating a new 'intermediate' geography to improve the allocation of immigrants to local areas, this has removed the tendency for over-estimation of immigration into London (ONS, 2007b). Despite the shortcomings of IPS/LFS data due to sample size, ONS believe that the data remain robust at the new intermediate geography (NMGi).

Analyses based on *NMD* data suggest an alternative methodology for distributing immigration flows which combines TIM statistics at a national level with sub-national statistics from three of the administrative sources already mentioned: HESA statistics, NINo registrations and GP registrations (Boden & Rees, 2010b). The methodology, which does not require the use of an intermediate geography, uses flow 'proportions' to distribute national TIM totals to sub-national areas and results in a redistribution of immigration flows that would have a significant impact upon mid-year estimates, sub-national projections and household projections in England and Wales. The impact of the alternative immigration estimation method upon the population of Leeds is illustrated in Fig. 6.7. The official statistics from ONS suggest that the population of Leeds has increased from 715,600 in 2001 to 761,100 in 2007. Simply replacing the immigration component of these annual estimates of population with the revised total calculated using the NMD results in a population for Leeds in 2007 of 734,300, almost 27,000 less than the published estimate.

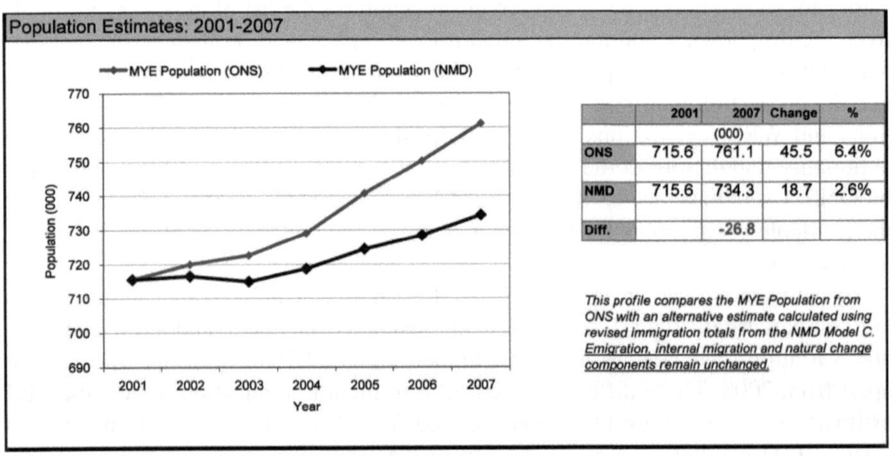

	2001	2007	Change	%
		(000)		
ONS	715.6	761.1	45.5	6.4%
NMD	715.6	734.3	18.7	2.6%
Diff.		-26.8		

Fig. 6.7 MYE population for Leeds: ONS population compared to NMD estimate (Sources: ONS mid-year population estimates; NMD model estimates)

In summary, the development of the *NMD* has exemplified not only the benefits of enhanced understanding of data from alternative sources relative to one another and therefore better monitoring of trends in immigration over time at local and regional level, but it has also demonstrated how methods making use of particular administrative data for estimating the international migration components of population change can identify potential inconsistencies in the mid-year population estimation methods that are so important for local resource allocation.

The European Demographic Databank (EDD)

In this section, we address issues that relate to the European dimension of migration statistics. DEMIFER (DEmographic and MIgratory Flows affecting European Regions and cities) is a project undertaken through the ESPON (European Spatial Planning Observation Network) 2013 Programme which is designed to reinforce EU regional policy (Objective 3 of Structural Funds 2007–2013) with studies, data and observations of development trends. The project is set in the context of the main demographic development in Europe in the next decades which is the ageing of the population that will cause a decline in the working age population and which may lead to a downward effect on economic growth and competitiveness in many European regions. The project also addresses the challenges that regions will face from environmental changes, particularly climate change and limitations in the availability of energy. The key aim of DEMIFER is therefore to assess the effects of demographic trends and migration flows on European regions and cities and to examine the implications for economic and social cohesion, taking into account the possible effects of climate change. The project therefore aims to answer the following types of questions: How will future developments in migration, fertility and mortality affect population growth and changes in the age structure in different types of regions? To what extent may the labour force increase due to increases in natural growth, internal migration, international migration and labour force participation rates in different types of regions? Which policy options may result in achieving increases in natural growth, migration and labour force participation? To what extent may the effects of internal migration, migration between European countries and migration to Europe compensate for each other or reinforce each other?

In order to address these types of questions, it was felt necessary to develop an up-to-date database, to distinguish between different types of migrants, to develop a typology of regions, to specify alternative scenarios and to use a multiregional demographic projection model to produce sets of projections. Research colleagues at the Nordic Centre for Spatial Development (Nordregio, Sweden) were responsible for the creation of a database of Eurostat's demographic statistics for use within the DEMIFER initiative. The *European Demographic Databank* (*EDD*) has been developed at the University of Leeds as an interface to these data providing NUTS2 summaries of the patterns and trends that exist in each of the key demographic components. The *EDD* contains time series data on births and deaths, as well as migrants between regions in the same country, migrants between countries and migrants entering Europe from the rest of the world. The European system involved the 27 member states of the EU plus four other countries: Iceland, Liechtenstein, Norway and Switzerland. The spatial units used across Europe are defined as Nomenclature of Statistical Territorial Units (NUTS) regions. There are currently 271 NUTS2 regions and 1,303 NUTS3 regions in the EU27 and there is no distinction between NUTS regions in Cyprus, Luxembourg and Liechtenstein. This section of the chapter focuses on the development of the information system with particular emphasis on the UK, the international migration flows required and the production of a time series of NUTS2 level inter-regional aggregate flow matrices for calendar years from 1999 to 2007 with accompanying NUTS2 region inflow, outflow counts, disaggregated by single year of age; progress on other parts of the project are available from the Interim Report (Bauer et al., 2009) or the draft final report (De Beer et al., 2010).

Initial examination of the Eurostat databases on migration indicated an absence of any data on internal migration within the UK at NUTS2 level. The spatial hierarchy of NUTS regions involves 12 NUTS1 regions (GORs), 34 NUTS2 regions (counties or groups of counties) and 133 NUTS3 regions (counties or groups of unitary authorities). In order to produce a time series of flows between NUTS2 regions in the UK, it was necessary to develop an estimation methodology and apply this to existing data. The estimation methodology used by Dennett and Rees (2010) can be summarised as follows:

- assemble PRDS data for England and Wales and Community Health Index (CHI) data for Scotland that capture the movements of NHS patients between districts within England and Wales and Scotland respectively and aggregate these flow matrices to NUTS2 regions;
- estimate the missing migration flows between NUTS 2 regions in England/Wales and in Scotland using NHSCR and 2001 Census data;
- estimate single year of age in, out and net flow data or NUTS2 regions using a combination of national single year of age migration profiles taken from the 2001 Census, single year of age population estimates produced by the different UK statistical agencies for constituent UK countries for each year of interest, and where available, migration estimates by broad age group from PRDS data; and
- adjust the data from mid-year to mid-year flows to calendar year flows.

After consultation with the Office for National Statistics, the new time series of internal migration data for the UK has been supplied to Eurostat as ‘experimental’ UK statistics. The internal migration data have also been incorporated with sets of data relating to the other components of population change in *EDD*, providing a facility to display a series of demographic indicators simultaneously for every NUTS2 region across the system of interest. Figure 6.8 exemplifies that part of the prototype system under construction that displays time series indicators of fertility, mortality and migration for the selected NUTS2 region of West Yorkshire. The top left graphs under fertility show the crude number of births in thousands from 1991 to 2006 and also expressed as an index set to 1 in 1991. The statistics illustrate the reversal of the decline in fertility that has occurred in West Yorkshire and which is

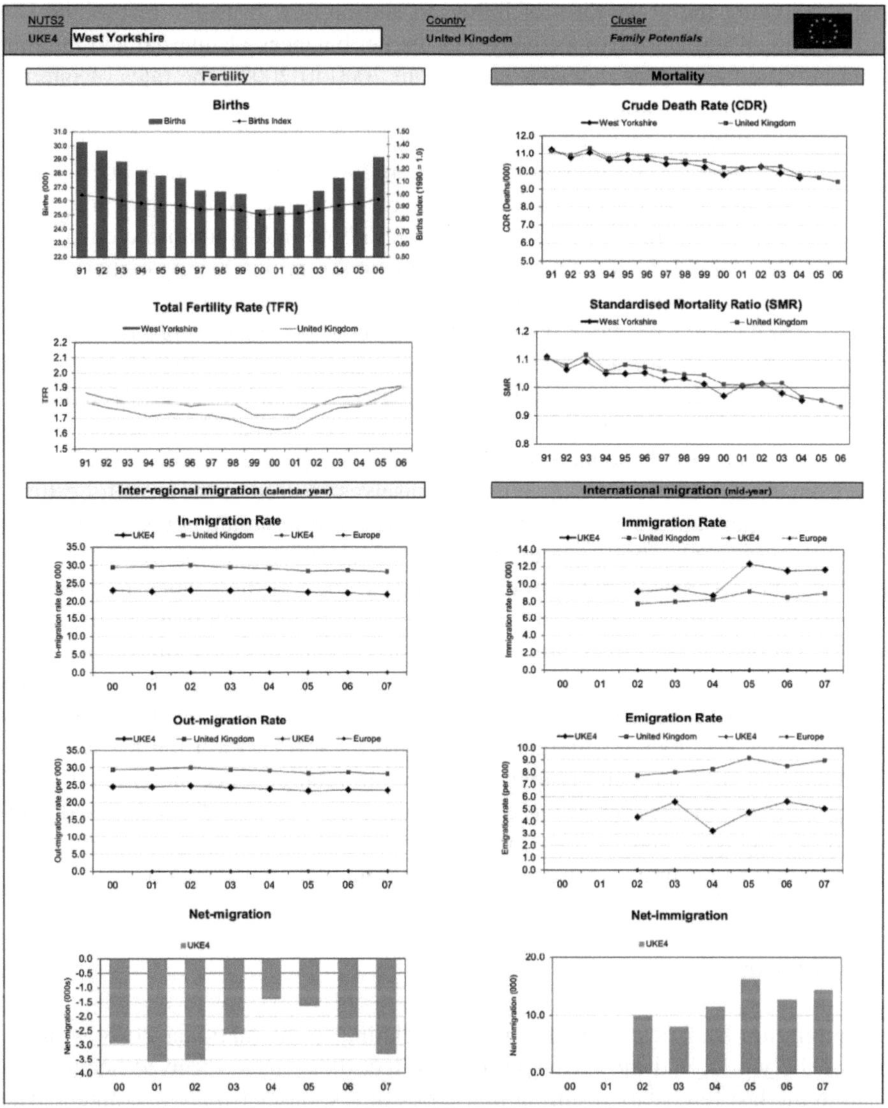

Fig. 6.8 *EDD* area profile: an example illustration for West Yorkshire (Sources: Fertility, mortality and migration data obtained from the Eurostat statistics database – http://epp.eurostat.ec.europa.eu/ – with some adjustment and estimation for error. Population at risk data obtained from ONS mid-year estimates – http://www.statistics.gov.uk/statbase/)

also evident from the total fertility rate graph. Two sets of data convey the declining trend in mortality over the same period. The crude death rate is below the national average in West Yorkshire but when mortality is standardised by the numbers of people in each age group, the standardised mortality ratios (SMRs) are above the national average in all years apart from 1998. The bottom panel of Fig. 6.8 displays graphs showing the gross rates of inflow to and outflow from the rest of the UK and the rest of the world respectively and the net balances of internal and international migration over the calendar years from 2000 to 2007. The rates of internal in-migration and out-migration for West Yorkshire are both below the national average and although they fall marginally during the period, the net loss in absolute terms declines by more than half between 2001–2002 and 2004–2005 before increasing again in 2006–2007. In contrast, West Yorkshire gains significantly from net immigration during the period with a peak in both rates and absolute net flows in 2005. The magnitude of the net gains are due to this region having higher rates of immigration and lower rates of emigration than the national average.

The *EED* system provides quick and convenient access to a profile of demographic components for any selected NUTS2 region in any of the 31 DEMIFER project countries. It allows comparison of patterns and trends in each region benchmarked against EU or national trends and therefore it enables variations across Europe to be identified and analysed. The system also provides quality assurance of data from a range of sources which may themselves be used as the basis for estimation where other data are missing or unknown. This is role should not be understated; there are many gaps – both spatial and temporal in nature – in the official Eurostat demographic data sets. Finally, in drawing this section of the chapter to a close, it is important to recognise that part of the successful implementation and use of the *EDD* is keeping it well maintained and populated with the most up-to-date information available. If this can be accomplished, then it becomes feasible to consider extending the system to a more disaggregate spatial scale such as NUTS3, or encompassing other key socio-economic data at NUTS2 level within the system so as to understand regional demographic dynamics in relation to economic and social change. Developments such as these might lead to the creation of a web-based, integrated demographic and socio-economic information and planning support system which would be of immense use to researchers and policy makers alike.

Conclusions

UK migration statistics, particularly those relating to international flows with the rest of the world, and the means used to collect them, have received heavy criticism in the last few years resulting in a series of responses from Government. The report of the Inter-departmental Task Force on Migration (2006) proposed a series of measures including: increasing sample sizes in surveys; collecting more data from migrants through landing cards; introducing a points-based system; developing the e-Borders project (which includes passport scanning); creating better links between the various data sources that are already used; and improving statistical and demographic models. It is too early to evaluate the success of these measures including the MSIP programme, since some of the measures will take several years to implement. The success of the e-Borders programme, which aims to record electronically all arrivals and departures (by scanning passports) will obviously depend on effective implementation of the technology but it also involves collecting detailed personal information from travel agents when flights or trains are booked as well as travel itineraries. The aim is to capture 60% of journeys out of the UK by the end of 2009 and 95% by the end of 2010. It will remain to be seen whether it will be possible to convert a set of person records of border crossings over periods of time to a set of statistics on long-term and short-term migrants and visitors. The migration information that may ultimately be extracted from the database is not seen as a replacement to the IPS, but as an additional set of statistics, yet there is no proposal at the moment to add any IPS-type questions to the e-Borders process and 'join up' these two methods for collecting migration statistics.

Without a population registration system, it is unrealistic to expect Government to collect comprehensive data on internal or international migration. However, it is possible, to explore other sources of migration data that will help to fill gaps in current knowledge (e.g. use of the School Census to measure migration children of school-age) and, as this chapter has shown, to develop systems that draw data from different sources together and use the data that exist to create estimates of data that are partial or missing or make adjustments to data to enable temporal comparison. *WICID* contains data sets derived from previous censuses, for example, that allow comparison with data for 2001 Census geographical areas and a huge amount of estimation work has been required in populating the internal and international migration sections of *EDD* alone. Analyses with the *NMD* data sets have shown how better mid-year population estimates can be produced with alternative methods of using administrative data. Estimates of the illegal resident population of the UK have been suggested (Pinkerton, McLaughaln, & Salt, 2004) based on an examination of data from periods of amnesty in other European countries and generating sub-national estimates of illegal migrants is certainly one challenge for the future.

The three examples of migration information systems that have been reported in this chapter have each been developed in academic research environments as limited-life projects funded by different organisations. It is appropriate to conclude with some comments about the future of each system. The *WICID* system is funded by the ESRC until 2011 and its future will depend on whether the Census Programme will continue to be supported beyond 2011 and whether access to the census interaction data will be deemed important. It appears that ONS are planning to release 2011 Census Special Migration and Workplace Statistics and new questions asked in the Census will generate new sets of migration data. However, the mechanism of data release from the census agencies is likely to involve new data feed technologies and consequently the *WICID* software will need to be adapted to accommodate the new data streaming methods. The flexibility of query-building that allows origin-destination data to be extracted for different geographical areas is a powerful user facility which, together with the analysis functionality that has been incorporated and the opportunity for users to access interaction data from previous censuses and from the PRDS/NHSCR for inter-censal years gives should enhance its chances of longevity.

The *NMD* has been developed at the University of Leeds as a prototype system and has influenced both the way in which ONS disseminates up-to-date information about international migration and the way in which the mid-year population estimates are being produced (Boden, 2009a, 2009b).The *NMD* has considerable potential for local application, providing intelligence on the most uncertain component of demographic change – international migration. A hybrid of the *NMD* has been produced for the Strategic Migration Partnerships of Yorkshire and Humber and the North West, combining a range of statistics on international migrants with asylum data from the UK Borders Agency.

If we consider the state of migration statistics in the UK to be poor, then the availability of migration data at national level, let alone regional level, across Europe, is nothing less than awful. There are many countries for which the information on migration held by Eurostat is partial or missing and for the data sets that do exist there are inconsistencies due to different conceptual, definitional and measurement differences used by different countries. The UK has defaulted on its requirement to provide matrices of migration between NUTS 2 regions but, through the DEMIFER project, a methodology has been developed for producing these estimates using data which are available in the public domain and, working in consultation with the ONS, these data have now been deposited with Eurostat and are available for others to use. The estimates created for the UK also now form part of a much larger European database of flows at the NUTS2 level across 31 states that we have called *EDD* and which represents an important basis on which to monitor and interpret trends, to conduct historical analyses of flow magnitudes and patterns and to develop evidence-based projections into the future.

The 1998 UN recommendations provide a common goal for improving international migration statistics and, ideally, those countries with population registers and administrative systems that record changes in the place of residence should be able to provide data on a reasonably consistent basis. There is, however, a need for real political will to make this happen, and real harmonisation will only be possible when national governments are prepared to take seriously the preparation of 'internationally comparable migration statistics' rather than simply seeking to improve national migration statistics. In the meantime, much can be achieved through the use of estimation methods and modelling techniques to generate consistent sets of international and internal migration. Building migration information systems that draw together alternative data sets is often the first and most important stage in this process.

Acknowledgements The authors acknowledge the influential role that Phil Rees has played in UK migration research and the major contributions he has made to the development of each of the systems exemplified in this chapter in particular. Each author has enjoyed fruitful collaborations with Phil over a number of years and learnt a great deal from his guidance and suggestions.

John Stillwell and Adam Dennett also acknowledge the support of ESRC through the Census Programme grant to the Centre for Interaction Data Estimation and Research (CIDER) (RES-348-25-0005) and the role of Oliver Duke-Williams in developing the *WICID* system. The UPTAP project award (RES-163-25-003) entitled 'What happens when international migrants settle' gave Peter Boden the opportunity to work on the *NMD*. The development of *EDD* has been supported by funding under the ESPON programme for the DEMIFER project which is being carried out by a team of researchers from the Netherlands Interdisciplinary Demographic Institute (NIDI, Netherlands), the University of Vienna (UNIVIE, Austria), the International Organization for Migration/Central European Forum for Migration and Population Research

(IOM/CEFMR, Poland), the School of Geography at the University of Leeds (United Kingdom), the Netherlands Environmental Assessment Agency (NEAA, Netherlands), the Nordic Centre for Spatial Development (Nordregio, Sweden), and the National Research Council (CNR, Italy).

References

Bauer, R., Brocket, S., De Beer, J., De Jong, A., Ekamper, P., Fassmann, H., et al. (2009). *DEMIFER: Demographic and migratory flows affecting European regions and cities, Applied research project 2013/1/3: Interim report.* The ESPON 2013 Programme, Netherlands Interdisciplinary Demographic Institute, The Hague.

Bilsborrow, R. E., Hugo, G., Oberai, A. S., & Zlotnik, H. (1997). *International migration statistics: Guidelines for improving data collection systems.* Geneva: ILO.

Boden, P. (2009a). *The new migrant databank.* Presentation at the UPTAP-GROS Scottish government workshop, Understanding population trends and processes, Thursday 12th February 2009, Edinburgh. http://www.geog.leeds.ac.uk/fileadmin/downloads/school/research/projects/migrants/GRO_UPTAP_Review_-_PB_-_Feb_2009.pdf

Boden, P. (2009b, September). *International migration – The local impact of uncertainty in national statistics.*, Paper presented at the 2009 British Society for population studies conference 2009, Brighton: University of Sussex. http://www.geog.leeds.ac.uk/fileadmin/downloads/school/people/research/p.boden/BSPSSept09.pdf

Boden, P., & Rees, P. (2010a). International migration: The estimation of immigration to local areas in England using administrative data sources. *Journal of the Royal Statistical Society, 173*(4), 707–731.

Boden, P., & Rees, P. (2010b). New migrant databank – Concept and development. In J. Stillwell, O. Duke-Williams, & A. Dennett (Eds.), *Technologies for migration and commuting analysis: Spatial interaction data applications* (pp. 111–132). Hershey, PA: IGI Global.

De Beer, J., Van der Gaag, N., van der Erf, R., Bauer, R., Fassmann, H., Kupiszewska, D., et al. (2010, April 29). *DEMIFER Demographic and migratory flows affecting European regions and cities applied research project 2013/1/3.* Draft Final Report, ESPON.

Dennett, A., Duke-Williams, O., & Stillwell, J. (2007). *Interaction datasets in the UK: An audit* (Working Paper 07/05). Leeds: School of Geography, University of Leeds.

Dennett, A., & Rees, P. (2010, Summer). Estimating a recent time series of internal migration flows fro the UK. *Population Trends, 140*, 82–105.

Dennett, A., & Stillwell, J. (2010). Internal migration in Britain, 2000–01, examined through an area classification framework. *Population, Space and Place, 16*(6), 517–538.

Hollis, J. (2008). *London borough migration: 2001–2006, DMAG Briefing 2008-10.* London: Data Management and Analysis Group, Greater London Authority.

House of Commons Treasury Committee. (2008). *Counting the population.* London: The Stationery Office.

House of Lords Select Committee on Economic Affairs. (2008). *The economic impact of immigration. Volume I: Report.* London: The Stationery Office.

Inter-departmental Task Force on Migration. (2006). *Report of the inter-departmental task force on migration.* London: National Statistics.

Jefferies, J., & Fulton, R. (2005). *Making a population estimate in England and Wales.* Office of National Statistics. http://www.statistics.gov.uk/Statbase/Product.asp?vlnk=575

NSCD. (2006). *Estimating international migration for population estimates – An information paper.* Titchfield: National Statistics Centre for Demography. http://www.gro-scotland.gov.uk/files/uk-international-migration-paper.pdf

ONS. (2003). International migration statistics. *National statistics quarterly review series report no 23.* London: ONS.

ONS. (2007a). *Estimating internal migration – Customer guidance notes.* Migration Statistics Unit, Office of National Statistics. http://www.ststistics.gov.uk/statbase/EXPODATA%5Ccommentary%5CEstimationinternalMigration.doc

ONS. (2007b). *Improved methods for estimating international migration – Geographical distribution of estimates of in-migration*. Migration Statistics Unit, Office of National Statistics.
ONS. (2008). *2006-based sub-national population projections for England – Methodology guide.* Titchfield: Office of National Statistics. http://www.statistics.gov.uk/downloads/theme_population/SNPP-2006_Methodology_Guide.pdf
ONS. (2009). *Methodology to estimate total international migration – 1991 to 2007.* Titchfield: Office for National Statistics.
Pinkerton, C., McLaughaln, G., & Salt, J. (2004). Sizing the illegally President population in the UK. Available at http://www.homeoffice.gov.uk/rds/pdfs04/rdsolr5804.pdf
Poulain, M., & Perrin, N. (2003, July). Can UN migration recommendations be met in Europe? *Migration Information Source.* Washington, DC: Migration Policy Institute (MPI). http://www.migrationinformation.org/Feature/display.cfm?id=139
Raymer, J. (2008). Obtaining an overall picture of population movement in the European Union. In J. Raymer & F. Willekens (Eds.), *Migration in Europe: Data, models and estimates* (pp. 209–234). Chicheester: Wiley.
Raymer, J., & Abel, G. (2008). *The MIMOSA model for estimating international migration flows in the European Union*. UNECE/Eurostat work session on migration statistics Working Paper 8. Geneva: UNECE/Eurostat.
Rees, P. (1977). The measurement of migration from census data and other sources. *Environment and Planning A, 9*, 247–272.
Rees, P., & Boden, P. (2006). *Estimating London/s new migrant population. Stage 1 – Review of methodology.* London: Major of London, Greater London Authority.
Rees, P., Stillwell, J., Boden, P., & Dennett, A. (2009). A review of migration statistics literature. In UK Statistics Authority, *Migration statistics: The Way Ahead?* (pp. 53–140). London: UKSA.
Statistics Commission. (2005). *Census and population estimates and the 2001 census in Westminster* (Final Report No 22). http://www.statscom.org.uk/C_403.aspx
Stillwell, J., & Duke-Williams, O. (2003). A new web-based interface to British census of population origin-destination statistics. *Environment and Planning A*, *35*(1), 113–132.
Stillwell, J., & Duke-Williams, O. (2006). Understanding the 2001 census interaction data: The impact of small cell adjustment and problems of comparison with 1991. *Journal of the Royal Statistical Society Series A*, *170*(2), 1–21.
Stillwell, J., Duke-Williams, O., & Dennett, A. (Eds.). (2010). *Technologies for migration and commuting analysis: Spatial interaction data applications*. Hershey: IGI Global (forthcoming).
Stillwell, J., Duke-Williams, O., Feng, Z., & Boyle, P. (2005). *Delivering census interaction data to the user: Data provision and software development* (Working Paper 05/01). Leeds: School of Geography, University of Leeds.
Stillwell, J., & Hussain, S. (2010). Exploring the ethnic dimension of internal migration in Great Britain using migration effectiveness and spatial connectivity. *Journal of Ethnic and Migration Studies*, *36*(9), 1381–1403.
Stillwell, J., Rees, P., & Boden, P. (Eds.). (1992). *Migration processes and patterns, Vol. 2: Population redistribution in the United Kingdom.* London: Belhaven Press.
Tyler, P., & Boden, P. (2010). Measuring and managing the impact of immigration. *The Yorkshire and Humber Regional Review*, *20*(1), 28–30.
UK Statistics Authority. (2009). *Migration statistics the way ahead?* London: UKSA.
United Nations. (1998). *Recommendations on statistics of international migration*. Department of Economic and Social Affairs, United Nations, New York, ST/ESA/STAT/SER.M/58/Rev.1.
Vickers, D., Rees, P., & Birkin, M. (2003). *A new classification of UK local authorities using 2001 census key statistics* (Working paper 03/3). Leeds: School of Geography, University of Leeds.

Chapter 7
Comparing Internal Migration Between Countries Using Courgeau's *k*

Martin Bell and Salut Muhidin

Introduction

Despite the acknowledged significance of migration as the pre-eminent component of population change, understanding of the way population mobility varies between countries is, as yet, poorly developed. One symptom is that measures of migration are conspicuous by their absence from tables of international statistical demographic indicators. While fertility, mortality, and even international migration now commonly appear in such lists, internal migration is invariably absent. The reasons for this omission are well established and include the multidimensional nature of the mobility process, differences in the way migration is measured, and problems of spatial and temporal comparability, all of which prejudice rigorous comparative analysis.

In responding to this deficit, analysts have focused on two main objectives: identifying the types of migration data collected in countries around the world, and establishing common metrics that can be used to compare mobility behaviour. For the former, the pioneering work is due to Rees and Kupiszewski (1996, 1999a) who took inventory of the data available in Europe. This work was later extended by Bell (2003, 2005) to cover the 191 member states of the United Nations. For the latter, the key proposals emanate from a joint Anglo-Australian project which identified 15 discrete indicators, in four main groups, that might be used to measure various facets of internal migration (Bell et al., 2002; Rees, Bell, Duke-Williams, & Blake, 2000). In parallel with these developments, attention has also been given to some of the specific problems of migration analysis, such as changes in statistical boundaries, which seriously prejudice temporal comparison of migration flows (see, for example, Blake et al., 2000; Boyle & Feng, 2002; Stillwell, Bell, Blake, Duke-Williams, & Rees, 2000).

While much of the groundwork would appear to have been laid, no concerted attempt has yet been made to apply this knowledge to generate substantive empirical

M. Bell (✉)
School of Geography, Planning and Environmental Management, University of Queensland, QLD, Australia
e-mail: martin.bell@uq.edu.au

J. Stillwell, M. Clarke (eds.), *Population Dynamics and Projection Methods*, Understanding Population Trends and Processes 4, DOI 10.1007/978-90-481-8930-4_7,

comparisons of population mobility. This is not to overlook the valuable contribution made by several recent studies (see, for example, World Bank, 2009; Vignoli, 2004; CEPAL, 2007), but there is a pressing need for more systematic research. Greater clarity is needed not only as to the extent of cross-national differences, but to establish their trajectory in a form that will aid policy development.

Building on recent work for the 2009 UN Human Development Report (Bell & Muhidin, 2009), this chapter pursues that goal by exploring one of the measures identified in the recent review by Bell et al. (2002), as originally proposed by Courgeau (1973a). The particular appeal of Courgeau's '*k*' statistic is that it purports to provide a single summary index of migration intensity which transcends the differences in zonal systems that commonly confound cross-national comparisons. We apply Courgeau's *k* to examine differences in mobility between 27 countries, using census data drawn primarily from the Integrated Public Use Microdata Series (IPUMS) database maintained by the University of Minnesota. Our goals are threefold: first, to establish the strengths and limitations of Courgeau's *k* as a summary measure of internal migration; second, to identify the extent of international differences in mobility; and third, to determine the general trajectory of internal migration over time.

We begin by reviewing prior attempts to compare mobility between countries and identify the key issues that confront analysis. The next section describes Courgeau's *k* statistic and sets out details of the IPUMS data. This is followed by our results, focusing first on cross-national differences, then temporal trends. We conclude by summarising the major substantive findings, appraising Courgeau's *k*, and identifying possible avenues for enhancement.

Prior Work

Prior interest in cross-national comparisons of internal migration has taken a number of forms. There are several collections which compare patterns and processes in different countries. A prominent example is the 'Handbook' assembled by Nam, Serow, and Sly (1990), which methodically describes the patterns of movement, selectivity, causes and consequences of migration in 21 countries dispersed widely around the world. Rees and Kupiszewski (1999a, 1999b) present a similar analysis focusing on 28 European countries. There are also specialized studies that draw comparisons between countries with respect to particular forms of internal migration, such as rural-urban migration (United Nations, 2000), counterurbanization (for example, Fielding, 1982; Champion, 1989) and return migration (for example, Newbold & Bell, 2001).

Others have been more concerned with the characteristics of internal migrants. Most prominent here is the work of Rogers and colleagues (see, for example, Rogers, Raquillet, & Castro, 1978; Rogers & Castro, 1981) who showed how model schedules could be fitted to the age profiles of migration, with a series of parameters summarising three key elements: the rise and fall in the migration curve, the height of the peaks and troughs, and their displacement along the age axis.

Rogers and Castro (1981) showed that for movement within countries the overall shape of the curve was generally invariant with geographic scale. However, particular types of move (for example, to and from capital cities) displayed distinctive profiles and comparison across 500 profiles drawn from seven countries revealed variation on several of the parameters reported, most notably the age at which peak migration occurred. Subsequently, Long (1992) extended the comparison to other demographic characteristics, showing how marital status composition and associated events contributed to differences in the age profile of migration between the United States, Great Britain, Japan and Ireland.

Underpinning both the patterns of movement and the characteristics of migrants, it is the overall level or intensity of migration that represents the fundamental basis for comparison between countries. An early pioneer in the field was Long (1991) who published what appears to be the first international 'league table' comparing countries with respect to mobility. Drawing on one and 5-year interval data from the 1980 round of censuses, Long (1991) reported crude migration intensities across 16 nations, revealing wide variations in the propensity to move, with high mobility in four new world countries (Australia, New Zealand, United States and Canada) and relatively low mobility prevalent across Europe. The scale of the differences identified by Long (1991) was striking, with the 1-year mobility rate in New Zealand three times that observed in Ireland.

More recently, the World Bank (2009) has published estimates of migration intensity drawn from household surveys for 35 countries, covering a range of dates between 1992 and 2005. Two sets of estimates are provided, representing migration among people of working age measured (i) as lifetime migration, and (ii) as recent migrations (over a 5-year interval). As with the data reported by Long (1991), the results point to substantial variation between countries. Lifetime labour mobility varied from 1.2% in Micronesia to 52.5% in Bosnia and Herzegovina while 5-year rates ranged from zero in Madagascar and Mozambique to 22.4% in Armenia. Unfortunately, details of the zonal geography underpinning these data are unclear: migrants are defined simply as individuals who were not living in the same 'district' in which they were born, or previously resided. Thus, the reported differences may be as much a product of differences in the zonal systems on which the data were collected rather than in the underlying propensity to migrate.

For the 23 countries in Latin America and the Caribbean, another recent set of cross-national comparisons of internal migration is included in the *Panorama Social de America Latina* (CEPAL, 2007), extending a range of data assembled previously by Vignoli (2004). As with the World Bank report, the CEPAL documents encompass both 5-year and lifetime migration estimates, but in this instance data are drawn from the census, and utilise the distinction between major and minor regions to provide two sets of migration intensities. As elsewhere, the results reveal remarkable variation, with lifetime migration intensities between major regions ranging from lows of 10–15% in Guatemala, Nicaragua and Bolivia to highs of 25–30% in Barbados, Antigua and Paraguay. For minor regions, the figures are substantially higher – peaking around 50% in the case of Chile. Intensities for 5-year migration intervals are correspondingly lower, but tend to display a similar rank order with

Antigua (13%) and Paraguay (11%) emerging as the most mobile countries, and Cuba, Nicaragua and Guatemala (all less than 3%) registering the lowest movement between major regions. As with the World Bank estimates, the problem with the CEPAL comparisons is that the zonal systems vary widely between countries: for example, the number of 'major regions' ranges from 7 (Provinces of Costa Rica) to 33 (Departments in Colombia), while 'minor regions' number 128 (Cantons in Ecuador) to 338 (Communes in Chile) and 1105 (Municipalities in Colombia).

Consistency between the migration intensities calculated by CEPAL (2009) and those presented by the World Bank (2009) would lend credibility to both sets of figures. Unfortunately, however, comparison across the 10 countries which appear on both listings reveals marked differences. Simple product-moment correlation coefficients between the two data sets generate coefficients of determination (r^2 values) close to zero, both for lifetime migration and for 5-year intensities. These inconsistencies and the general dearth of comparative studies of internal migration can be traced to a range of impediments.

Impediments to Cross-National Comparison

Bell et al. (2002) identify four broad groups of problems that hinder comparisons between countries with respect to internal migration. These derive from differences in (i) the way migration is measured, (ii) population coverage, (iii) the intervals over which it is measured, and (iv) the division of space. We briefly consider each in turn.

Migration can be measured in several ways but the two most common methods either capture migration as an *event*, or detect it as a *transition* between two points in time. Migration events are generally associated with population registers and administrative datasets which record discrete changes of address. Population censuses, on the other hand, generally record transitions, measuring migration by comparison of a change in residence between two points in time. Over short observation spans (e.g. 1 year), counts of events broadly approximate recorded transitions, but as the observation interval lengthens, the two measures diverge at an increasing rate because the transition data fail to capture a growing incidence of repeat and return moves (Long & Boertlein, 1990). The two data types also differ in age-time plans which seriously compromises comparison of age schedules (Bell & Rees, 2006). Difficulties are avoided provided attention is confined to a single data type, but as more countries dispense with formal censuses (Langevin, Begeot, & Pearce, 1992), the problems of assembling a global inventory of comparable indicators are set to increase.

Differences also occur in population coverage and in the way particular sub-groups are dealt with in regard to migration statistics. While census under-enumeration is low in most countries, it is invariably highest among the most mobile members of the community. In the 1991 Australian census, for example, the rate of under-enumeration at 1.9% was comparatively low, but reached a high of 16.1% among people who were enumerated away from home (Australian Bureau of Statistics, 1995). At the same time, countries differ in the way they deal with groups such as the armed forces, diplomats, guest workers, overseas visitors, domestic and foreign students, the homeless and those with no fixed abode – all of whom typically display high mobility. Even definitions of usual residence vary while some countries still make comparisons based on place of enumeration (United Nations, 2008). Registration data, too, commonly exclude some segments of the population.

Another key factor that confounds comparisons is variation in the interval over which migration is observed. Three basic approaches can be identified, distinguishing countries which compare place of residence at the time of the census with: place of birth; place of residence at some fixed point in the past (typically 1 or 5 years ago); and place of previous residence, irrespective of when the move occurred. Each has advantages. Transitions measured over a fixed interval are most straightforward to analyse but different intervals carry particular benefits: data measured over a single year best reflect respondent characteristics at the time of migration, and hence are most effective in capturing migrant selectivity; 5-year data best reflect patterns of redistribution, free from the influence of short-term effects, but risk greater errors in recall and suffer higher non-response. Birthplace data reveal lifetime migrations which reflect long-run patterns of redistribution, but miss a large volume of intermediate moves. Because they inherit cumulative movements over the full lifespan, cross-national comparisons are also distorted by age composition effects. Place of previous residence data feature less commonly in national collections, though they are potentially rich sources of information when coupled with fine-grained data on duration of residence (Xu-Doeve, 2006).

These distinctions are important because comparing migration measures having different observation intervals can produce misleading results. Multiplying the number of migrants captured in a 1-year transition interval by five does not provide a reliable estimate of the number of migrants during a 5 year transition interval, because an increasing proportion of moves is made by 'chronic migrants' (Courgeau, 1973b; Kitsul & Philipov, 1981; Long & Boertlein, 1990; Rogerson, 1990). In a similar way, lifetime migration represents the cumulative effect of multiple moves and lifetime intensities are generally higher than fixed interval measures, but not by a fixed ratio. Differing migration intervals also reveal differing patterns of population redistribution (Rees, 1977).

Which of these migration intervals is to be preferred for cross-national comparisons? In practice, the choice of measure is dictated by the data that are most commonly collected. Bell (2005) found that of the 191 UN member states, 141 collected some form of migration data at the census. Of these, birthplace data were the most common, being collected by 115 nations; 94 measured migration over a fixed interval, with the 5-year timespan being the most common, while 34 collected data on place of previous residence but with no reference date. Several countries collected more than one form of data. Table 7.1 indicates that 5-year data were most common in Asia and Latin America, while the 1-year interval was confined primarily to European countries. Place of birth data featured across all continents but were least ubiquitous in Asia.

In the spatial domain, it is differences in statistical geography that prejudice comparisons, because the number of migrants recorded in any form of data collection is fundamentally dependent on the number and shape of the units into which

Table 7.1 Countries collecting transition data at the census by continent and data type

Continent	Place of birth	Defined interval: One year	Five years	Other defined interval	No reference date	Total countries
Africa	32	6	7	10	10	38
Asia	22	3	16	4	12	35
Europe	23	14	5	11	4	26
Latin America	23	1	17	3	6	27
North America	2	1	2	0	0	2
Oceania	13	3	9	1	2	13
TOTAL	115	28	56	29	34	141

Source: Adapted from Bell (2005).

a territory is divided. These issues, commonly grouped under the rubric of the Modifiable Areal Unit Problem (MAUP), affect all spatial comparisons but are especially pertinent to the analysis of migration (Bell et al., 2002; Wrigley, Holt, Steel, & Tranmer, 1996). Migration data are generally collected by reference to place of residence at two points in time, with the response coded to zones based on administrative or statistical divisions of a country. However, nations vary widely in size and shape, and in the geography of zonation (Law, 1999). In some countries, previous residence is coded to an individual village or town, while in others, only state or province of prior residence is recorded. Clearly comparisons which set migration intensities for countries based on differing numbers of zones are potentially misleading.

Analysts have proposed a number of possible solutions to this problem. One approach is to only compare countries with respect to *all* moves, rather than confining attention to only that subset of moves which cross zone boundaries. This is the approach adopted in the early work by Long (1991) cited earlier. In practice, however, relatively few countries collect such data. Bell (2005) identified just 37 countries for which it was possible to compute a migration intensity which included all moves, and for most cases this relied on data by duration of residence. Another approach is to develop a broadly comparable set of regions in each country, based around some form of functional classification. For example, Stillwell et al. (2000) used a framework of some 35–40 city regions to compare migration effectiveness in Britain and Australia, based around a functional classification of space that distinguished a set of metropolitan cores and their 'tributary' hinterland areas (rest, near, coast, far and remote) which organize the spatial economic systems in the two countries. While this proved an effective strategy for examining similarities and differences in two countries, it is less well suited to multiple comparisons. In an analysis for countries in Latin America and the Caribbean, Vignoli (2004) took an intermediate approach by distinguishing migration according two levels of geographic scale – those occurring between 'major' regions, such as states or provinces, and those between 'minor' regions such as communes or municipalities (see also CEPAL, 2007). While this avoids gross distortions, it goes only part way to addressing the MAUP since the number of zones in each category still varies widely from one country to the next.

Towards More Rigorous Comparison – Courgeau's '*k*'

Taken together, these considerations constitute a significant obstruction to the development of comparable indicators of internal migration. In general, differences between countries in population coverage can only be noted and an attempt made to assess their effects. Differences in data types and migration intervals are less tractable, and essentially constitute major divisions across which rigorous comparisons cannot reliably be made. Spatial issues have typically presented the thorniest problem, since no two countries have the same zonal system. Paradoxically, however, it is these very differences in zonation that open an intriguing avenue towards a comparative statistic.

Drawing on the ideas proposed by Courgeau (1973a), the aim is to create a synthetic indicator of migration for each country by coupling migration intensities at a range of spatial scales. As Courgeau (1973a) observed, if, as we know to be true, there is a relationship between the propensity to move and distance, there must also be a relationship between the level of mobility and the number of zones into which a space is divided. The finer the spatial mesh, the larger number of migrations that will be recorded, and hence the greater the apparent migration intensity. Courgeau's (1973a) formula:

$$\mathrm{CMI} = k \ln(\mathrm{n}^2) \tag{7.1}$$

endeavours to capture this link in a simple linear equation which connects the crude migration intensity (CMI) to the log of the number of regions (n) squared. [1] For any system of zones, the value of *k* can be estimated by calculating the CMI at various levels of the spatial hierarchy, each corresponding to a particular n. The CMI for a given n is calculated as the total number of migrations which cross the boundaries between zones, normally expressed as a percentage of the population at risk (see Rees et al., 2000): movements within zones at each tier in the hierarchy are ignored. The resulting parameter *k* represents the slope of a simple regression line which ties together the observations of migration intensity at various levels of spatial disaggregation. Differences in the value of *k* then effectively capture variations between countries in the overall level of migration intensity – the higher the value of k, the greater the intensity of migration and *vice versa*. While Courgeau's *k* has no intrinsic, plain language meaning, it serves to capture the extent of the difference in migration intensity between countries which cannot be discerned from the raw data because of differences in their zonal systems.

As Courgeau (1973a) points out, the hypothesised relationship makes a number of simplifying assumptions with respect to the shape of the country, the size

---

[1] Courgeau (1973a) variously uses log(n) and log (n^2) in his theoretical exposition and subsequent analyses. For the analysis reported here we utilise the squared form of the equation.

and shape of the zones, the density of population and the frictional effects of distance. When applied to different countries, these assumptions are unlikely to be borne out in practice. Nevertheless, empirical testing delivered encouraging results. Courgeau's sample of 11 countries varied widely in respect of observation intervals so they could not be compared directly. However, within countries, Courgeau (1973a, table 1) found that Equation (7.1) delivered a remarkably stable figure for k, implying a linear relationship between the two variables. For the three countries with a comparable (5 year) migration interval, the USA, England and France (adjusted to 5 years by Courgeau), and using data for the first half of the 1960s, he reported k values of 5.6, 4.5 and 4 respectively, which are broadly consistent with contemporary understanding of cross national differences: migration intensity in the USA is substantially higher than in most European countries (Long, 1991). How well does this relationship hold across a wider range of countries?

The IPUMS Data

Apart from the technical challenges involved, one of the fundamental impediments to cross-national comparisons has been the dearth of available data. Few nations make internal migration statistics readily available in electronic form or in standard reports, and there is no central statistical agency that forms a repository for such data. However, one source which provides information on internal migration for more than one country is the IPUMS database, maintained and made publicly available by the University of Minnesota. At the time of our analysis, 35 countries were represented in the IPUMS database, though not all of these collect data on internal migration. Moreover, because the IPUMS data are public use sample files, the available datasets exclude some census variables, or have limited classificatory detail. In the endeavour to preserve confidentiality, geographic attributes are often abbreviated, or limited to the regional level. Despite these constraints, the IPUMS database represents a unique resource for cross-national comparisons of mobility.

We draw on the IPUMS data for 25 countries. Information for two additional countries, Australia, and Indonesia, was held separately by the authors, while data for India and for the 2000 Census of China was kindly made available by colleagues via the United Nations. Our dataset therefore encompasses 28 countries in all, of which five are from Africa, eight from Asia, nine from Latin America and the Caribbean, and six from the developed world.

Table 7.2 sets out the date of the most recent census for which we have information for each country, together with the time interval over which migration was measured. It is readily apparent that a large number of countries collect more than one type of data, but the most common intervals over which migration is measured are 5 years (19 countries) and lifetime (25 countries). Six countries collect data for a 1-year interval and two collect data for a 10-year interval. Nine countries also collect data on place of previous residence, irrespective of the time of the last move.

Table 7.2 Measures of internal migration from the census, selected countries

		Internal migration interval				
Country	Census year	One year	Five years	Ten years	No reference date	Lifetime
Africa						
Ghana	2000		x			X
Kenya	1999	x				x
Rwanda	2001				x	x
South Africa	2002		x			x
Uganda	2001				x	x
Asia						
Belarus	1999				x	x
Cambodia	1998				x	x
China*	2000		x			x
India*	2001				x	x
Indonesia*	2000		x		x	x
Malaysia	2000		x			x
Philippines	2000		x	x		x[a]
Vietnam	1999		x			
Latin America and Caribbean						
Argentina	2001		x			x
Brazil	2000		x			x
Chile	2002		x			x
Colombia	2005		x		x	x
Costa Rica	2000		x			x
Ecuador	2001		x			x
Mexico	2005		x			x[b]
Panama	2000				x	x
Venezuela	2001		x			x
Developed countries						
Australia*	2006	x	x			
Canada	2001	x	x			x[c]
Portugal	2001	x	x			x
Spain	2001			x	x	x
UK	2001	x				
USA	2005		x[b]			x

*Data are provided from sources other than IPUMS.
[a]Only available from the 1990 Census.
[b]Only available from the 2000 Census.
[c]Collected but not useable in the IPUMS data.
Source: Abbreviated from Bell and Muhidin (2009).

For the purposes of this analysis, we confine attention to the two categories which provide the largest samples: those with the 1-year fixed interval and lifetime data. Together, these two intervals cover 27 of the 28 countries in Table 7.3; we omit the UK from further analysis.

Table 7.3 Five-year migration intensity by country and zonal system

	Country	Zonal system	No. of zones	Migrants	Intensity (%)
Africa					
	Ghana	Region	10	567,590	3.52
		District	110	961,270	5.96
	South Africa	Province	9	1,704,363	4.26
		Municipality	52	5,275,618	13.18
Asia					
	China	Province	31	32,347,800	2.74
		County	2,901	79,052,151	6.70
	Indonesia	Region	7	1,507,406	0.83
		Province	26	3,954,104	2.19
		Municipality	280	6,917,713	3.90
		Municipality	314	7,089,722	3.98
	Malaysia	State	15	840,800	4.75
		District	133	1,395,950	8.00
	Philippines	Region	16	1,559,511	2.51
		Province	83	2,038,365	3.28
		Municipality	1,610	2,823,789	4.55
	Vietnam	Region	8	1,337,724	1.94
		Province	61	1,999,215	2.90
		District	663	3,139,252	4.55
		Commune	1,203	4,481,825	6.50
Latin America and Caribbean					
	Argentina	Province	24	1,161,800	3.55
		Department	511	2,358,080	7.21
	Brazil	Region	5	3,372,124	2.20
		State	27	5,204,886	3.40
		Municipality	1,520	15,314,989	9.99
	Chile	Region	13	853,960	6.32
		Province	44	1,295,150	9.59
		Municipality	178	2,253,170	16.68
	Colombia	Department	33	1,520,980	4.21
		Municipality	532	2,302,190	6.42
		Municipality	1,105	2,676,375	7.39
	Costa Rica	Province	7	184,260	5.53
		Canton	60	353,010	10.60
		Canton	81	355,220	10.67
	Ecuador	Province	22	595,020	5.55
		Canton	128	885,170	8.25
	Mexico	State	32	2,470,960	2.70
	Venezuela	State	24	1,022,660	5.07
Developed countries					
	Australia	State/Territory	8	779,951	4.76
		Stat. Division	61	1,689,879	10.39
	Canada	Province	11	908,962	3.37
		Census.Division	288	3,359,319	12.46
		Municipality	5,600	4,466,827	16.57
	Portugal	Region	7	183,340	1.92
		Sub Region	22	307,940	3.23
		Municipality	308	677,380	7.10

Table 7.3 (continued)

Country	Zonal system	No. of zones	Migrants	Intensity (%)
USA	Parish	4,000	1,374,960	14.42
	Region	4	12,243,724	4.80
	Division	9	16,740,835	6.57
	State	51	22,794,783	8.94

Source: Modified after Bell and Muhidin (2009).

Comparing Migration Intensities

Tables 7.3 and 7.4 set out internal migration intensities for the various countries for which data are available, using each country's native statistical geography. In some cases (for example, Mexico and Venezuela), data are only available for a single level of geography, or zonal system, such as state or province, whereas for other countries, Brazil and Chile for example, the data are coded to several geographic levels. There are some cases, too, where migration data are available for alternative aggregations of the same type of zonal units. In the case of Colombia, for example, the IPUMS dataset includes a 'migration status' variable that indicates the proportion of people who changed residence between the country's 1,104 municipalities over the 5-year transition interval. This variable captures the aggregate migration intensity, but provides no information on spatial patterns. However, the IPUMS data set also provides an origin-destination matrix showing flows between 532 zones which are similarly named (i.e. as municipalities), but involve aggregations of zones with smaller populations. Because they involve different spatial breakdowns, the two variables deliver quite different estimates of migration intensity.

The results point to variations between countries in migration intensity. In the case of fixed interval data, Table 7.3 reveals a low of less than 0.8% of people moving between the eight regions of Indonesia over the previous 5 years, whereas 16.7% of people relocated between the 178 municipalities of Chile over the same period. High movement intensities, around 10% or more, were also recorded between municipalities in South Africa, Canada and Brazil, between parishes of Portugal, cantons of Costa Rica, and statistical divisions of Australia. In contrast, regions of Vietnam and Portugal registered intensities of less than 2%. In terms of absolute numbers, China, the USA, Brazil, Indonesia and South Africa stand out, each with more than 5 million people relocating between geographic zones over the 5-year interval.

The figures for lifetime migration intensity are consistently higher than for the 5-year period and in several cases reveal a remarkable level of mobility. Thus, in Chile, an astonishing 50% of the population were living outside their municipality of birth by the time of the 2002 Census. The same was true for two-fifths of Brazilians and Spaniards, and for one third of Colombians, Panamanians (districts) and Costa Ricans (cantons). Lifetime migration was much less common between the States of India, or between provinces of China and of Indonesia, with intensities of less

Table 7.4 Lifetime migration intensity by country and zonal system

	Country	Zonal system	No. of zones	Migrants	Intensity (%)
Africa					
	Ghana	Region	10	3,329,320	17.75
		District	110	5,206,990	27.75
	Kenya	Province	8	3,496,560	12.64
		District	69	5,622,520	20.32
	Rwanda	Province	12	801,890	10.41
	South Africa	Province	9	6,717,270	15.36
	Uganda	Region	4	1,288,730	5.24
		District	56	3,577,610	14.56
Asia					
	Belarus	Region	6	944,270	10.78
	Cambodia	Province	24	1,308,780	11.65
		District	149	2,024,170	18.02
	China	Province	31	73,087,300	6.19
	India	State	35	42,341,703	4.14
		District	593	76,841,466	7.52
	Indonesia	Region	7	8,104,818	4.07
		Province	26	16,729,095	8.39
	Malaysia	State	15	4,156,500	20.71
	Philippines	Region	16	6,879,231	11.72
		Province	77	8,722,805	14.86
Latin America and Caribbean					
	Argentina	Province	24	6,691,210	19.90
	Brazil	Region	5	17,025,306	10.07
		State	27	26,059,033	15.41
		Municipality	1,520	63,461,867	37.52
	Chile	Region	13	3,097,070	21.27
		Province	44	4,324,420	29.71
		Municipality	338	7,258,850	49.61
	Colombia	Department	33	8,108,168	20.25
		Municipality	532	12,452,428	32.51
		Municipality	1,105	14,589,440	36.23
	Costa Rica	Province	7	704,020	20.02
		Canton	60	1,195,490	33.99
		Canton	81	1,203,560	34.22
	Ecuador	Province	22	2,431,310	20.23
		Canton	128	3,641,200	30.30
	Mexico	State	32	17,791,208	18.52
	Panama	Province	11	566,940	20.56
		District	75	950,050	34.46
	Venezuela	State	24	5,184,850	23.79
Developed countries					
	Portugal	Region	7	1,240,580	12.80
		Sub Region	22	1,817,780	18.76
	Spain	Province	52	8,641,300	22.36
		Municipality	366	17,288,760	44.75
	USA	Region	4	44,423,142	17.84
		Division	9	57,909,783	23.25
		State	51	78,583,779	31.55

Source: Modified after Bell and Muhidin (2009).

than 10%. Nevertheless, the absolute numbers living outside their district, province or region of birth in the more populous countries was substantial: 78 million in the USA (states), 73 million in China (provinces), 63 million in Brazil (municipalities) and 42 million in India (states). In each case this implies a substantial historical shift in the pattern of human settlement.

There is a strong linear correlation between the levels of migration intensity measured over 5 years with that measured over the entire lifetime. Comparing Tables 7.3 and 7.4, there are 30 cases in which we have estimates of migration intensity for the same countries and geographic levels. The product moment correlation (Pearson r) between these two indicators generates a coefficient of determination (r^2) of 0.87. Thus, fixed interval intensities represent a reasonably reliable surrogate for lifetime moves.

What is most striking from these tables is the extent to which the magnitude of the computed intensities is dependent upon the level of spatial disaggregation at which migration is measured. This fundamentally undermines any simple comparison of migration intensities. For example, South Africa appears to be a middle ranking nation if migration intensity is measured between provinces (4.26% over 5 years), but ranks near the top of the list if the intensity is computed between municipalities (13.18%). Distinguishing between two levels of geography (major regions and minor regions) as in CEPAL (2007) only goes partway to addressing the problem. For movements between States, Malaysia recorded a higher 5-year intensity than South Africa (4.75% compared with 4.26%), but Malaysia has 15 states compared with South Africa's nine provinces, so it is unclear whether the higher intensity reflects greater underlying population mobility or if it is simply an artefact of the more disaggregated zonal system. Ecuador's migration intensity of 5.5% between 22 provinces raises a similar conundrum, while for minor regions, it is unclear whether movement between municipalities is greater in Chile (16.7%) than in South Africa (13.2%) because of higher underlying mobility, or because of the larger number of zones over which it is measured (178 compared with 52).

Cross-National Comparison Using Courgeau's '*k*'

Figure 7.1 plots the observed 5-year intensities from Table 7.3 against the natural logarithm of the corresponding number of regions squared, following the relationship proposed by Courgeau (1973a). The results reveal an intriguing picture. For 10 of the 19 countries, we have data for three or more zonal systems and in each case the graph reveals that migration intensity is approximately a linear function of n^2. Moreover, in the seven countries with just two data points, the slope of the line more or less matches those of the former countries. In each case, the intercept is oriented towards the bottom left corner of the graph, and the general impression is of a fan-like arrangement emanating from the origin.

The graph suggests that the linear equations for each country can be read as a continuous function, thereby allowing comparisons to be drawn at any chosen level of disaggregation (certainly by interpolation, if not by extension). Thus,

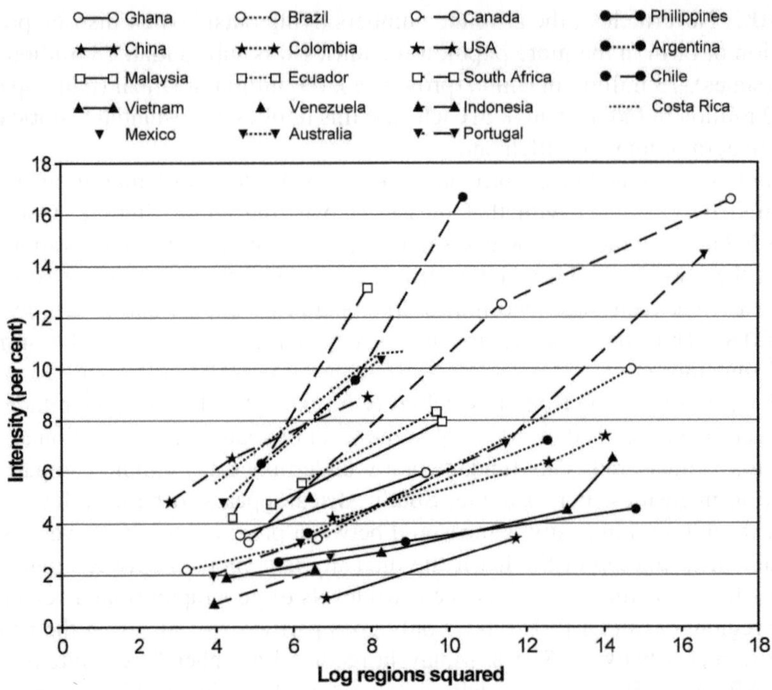

Fig. 7.1 Five-year migration intensity by zonal system, selected countries (Source: Modified after Bell & Muhidin, 2009)

disregarding differences in the original zonal systems against which the data were collected, comparisons could be made for each country on a system of, say, 10 or 20 regions, by reading the intensity on the y-axis for any chosen spatial level on the x-axis, or substituting n=x in the appropriate equation.

According to Courgeau's hypothesis, the key statistic summarising these associations is the slope of the regression line, captured in the index, *k*. Table 7.5 sets out the values of *k* for the 17 sample countries, with the regression line forced through the origin and the coefficient of determination reported as a measure of goodness of fit. The results indicate a close linear association in all except two cases, the Philippines and the USA, where r^2 falls below 0.8. This analysis greatly clarifies the mass of individual intensities reported in Table 7.3. According to the *k* values, South Africa and Chile emerge as the most mobile countries, followed closely by Costa Rica, the USA and Australia. These are followed by a second group, led by Canada, but drawn from widely around the world, comprising Ecuador, Malaysia and Portugal. Next in line come three Latin American countries, accompanied by Ghana, all displaying moderate levels of mobility, and these are followed by four Asian representatives, Vietnam, the Philippines, Indonesia and China, forming a distinctive group with the lowest levels of migration intensity.

While there is no discrete ordering of internal migration propensity by world region or level of development, some clustering is apparent from Fig. 7.1 and

Table 7.5 Courgeau's k and r^2 for selected countries, 5-year migration interval

	Country	k	r^2	Floating intercept
Africa				
	Ghana	0.660	0.903	1.176
	South Africa	1.503	0.819	−6.918
Asia				
	China	0.432	0.994	−2.10
	Indonesia	0.338	0.963	−0.621
	Malaysia	0.828	0.988	0.689
	Philippines	0.337	0.674	1.310
	Vietnam	0.403	0.890	−0.081
Latin America and Caribbean				
	Argentina	0.574	0.998	0.242
	Brazil	0.654	0.972	−0.556
	Chile	1.455	0.895	−4.456
	Colombia	0.530	0.942	1.107
	Costa Rica	1.268	0.964	1.303
	Ecuador	0.864	0.984	0.799
Developed countries				
	Australia	1.239	0.988	−1.004
	Canada	0.987	0.959	−1.042
	Portugal	0.757	0.904	−2.593
	USA	1.267	0.570	2.800

Source: Modified after Bell and Muhidin (2009).

Table 7.5. Led by Australia and the USA, the four developed countries all display relatively high values of k, but they are eclipsed by several developing countries, particularly South Africa and Chile. Latin American countries also feature prominently in the upper reaches of the chart, but there are stark contrasts between high mobility in Chile and Costa Rica, and comparatively subdued rates of movement in Brazil, Argentina and Colombia. Ecuador assumes an intermediate position. Asian countries generally display lower mobility, but Malaysia is a noticeable exception with a k value double that of its fellow countries, roughly equivalent to the mean of the Latin American cluster. African countries are poorly represented in this sample, but South Africa and Ghana display radically different profiles, suggesting very different migratory environments.

Graphing the underlying values, as in Fig. 7.1, also enables those countries to be located for which only a single data point is available, corresponding to the intensity computed for a single zonal system. For the countries in Table 7.3, this is the case only for Venezuela and Mexico. For the former, its position on the chart suggests a moderate level of mobility, similar to Ecuador and Malaysia. For the latter, the intensity of movement between 32 states places it much lower on the graph, closer to the Asian countries than to its Latin American counterparts.

Figure 7.2 and Table 7.6 provide corresponding information for those countries which collect lifetime migration data. In this case, there are 16 countries for which intensities are available for two or more zonal levels, and six countries for

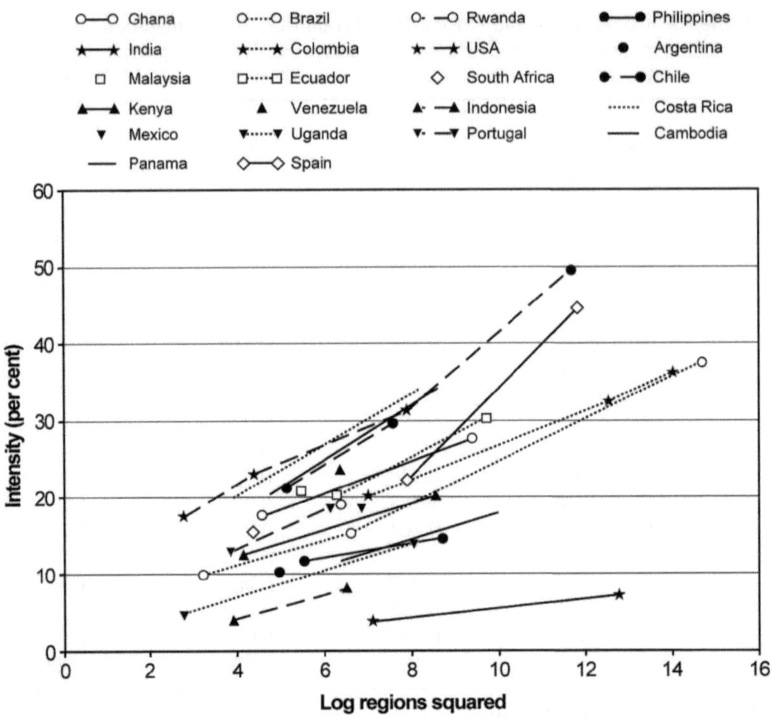

Fig. 7.2 Lifetime migration intensity by zonal system, selected countries (Source: Modified after Bell & Muhidin, 2009)

which there is just a single observation (Table 7.4). As might be expected, there is a close correspondence in the relative positioning of countries on the lifetime measure with that observed for the 5-year intensities, though in absolute terms the lifetime indices are consistently higher. Calculated across the 10 countries with k values for both series, the coefficient of determination (r^2), stands at 0.895. The broad sequencing of geographic regions is also maintained, with Latin American countries occupying the upper reaches of the graph, and Asian nations positioned in the lower margins. Among the developed countries, the USA stands out with the highest overall k index, while Portugal and Spain both feature in the middle ranks. The African representatives, once again, are widely scattered.

Figures 7.1 and 7.2 suggest that the linear equations for each country can be read as a continuous function, thereby allowing comparisons to be drawn at any chosen level of spatial disaggregation, not only by interpolation between observed values, but also by extension. Thus, disregarding differences in the original zonal systems against which the data were collected, calculations could be made to compare the implied level of migration intensity for each country on a system of, say, 10 or 20 regions by reading off the intensity on the y-axis for any chosen point on the x-axis, or by simply substituting the number of regions in the appropriate equation. However, it is apparent that the country-specific graphs in Fig. 7.1 are not perfectly

Table 7.6 Courgeau's *k* and r^2 for selected countries, lifetime migration

	Country	*k*	r^2	Floating intercept
Africa				
	Ghana	3.127	0.723	8.135
	Kenya	2.524	0.807	5.223
	Uganda	1.817	0.999	0.352
Asia				
	Cambodia	1.553	0.999	0.391
	India	0.587	0.999	−0.098
	Indonesia	1.318	0.995	−2.338
	Philippines	1.827	0.280	6.177
Latin America and Caribbean				
	Brazil	2.543	0.987	0.938
	Chile	4.159	0.989	−2.200
	Colombia	2.625	0.971	4.434
	Costa Rica	4.121	0.850	8.326
	Ecuador	3.153	0.994	1.841
	Panama	4.071	0.985	3.083
Developed countries				
	Portugal	3.107	0.960	2.681
	Spain	3.493	0.841	−22.953
	USA	4.500	0.418	10.953

Source: Modified after Bell and Muhidin (2009).

aligned towards the origin – numerous crossovers are apparent, so the apparent relativity between countries would depend upon the level of disaggregation at which migration intensity is measured.

It is also apparent that, in most cases, fitting simple linear equations to the available observations for each country would result in a non-zero intercept on the y-axis. Other things being equal, positive intercepts imply a greater tendency towards long-distance migration than might otherwise be expected (since they are driven by higher intensities over smaller levels of disaggregation), whereas negative intercepts suggest that migration over long distances is less prevalent than expected, given the level of mobility over shorter distances. As shown in Tables 7.5 and 7.6, positive intercepts are most pronounced in the case of the USA, which is consistent with a surfeit of long-distance movement compared with other countries. Ghana, Kenya, the Philippines, Columbia and Costa Rica also display relatively high positive intercepts. Negative intercepts are more prevalent in the 5-year data, and are largest in the case of South Africa, China, Portugal and Chile, suggesting that, in these countries, short-distance moves tend to predominate. In the case of lifetime measures, however, it is Spain that delivers the largest negative intercept, symptomatic of relatively low lifetime redistribution over longer distances, if taken at face value. In practice, though, it can be argued that the function linking migration intensity to distance in any country must, in reality, pass through the origin, because when the nation is reduced to a single region, internal migration, by definition, reduces to zero. This, in turn, suggests that the nature of the distance-decay relationship actually varies between countries, and rarely corresponds to the simple power function assumed for the analysis presented here.

Trends in Internal Migration

Is the propensity to change residence within countries rising or falling? It is widely asserted that mobility rose strongly over the course of the twentieth century, but the evidence is sparse, fragmented and inconclusive. In the United States, there is clear evidence that aggregate mobility rates have been falling steadily since the 1960s (Bogue, Ligel, & Kozloski, 2009). In China, on the other hand, loosening of restrictions has underpinned rapid growth in rural to urban migration (Fang, Du, & Meiwan, 2009). Courgeau's *k* provides a useful index to explore this question across a number of countries, free from the influence of the frequent changes in statistical boundaries which often occur from one census to the next.

Figures 7.3 and 7.4 illustrate the trends for our sub-set of countries, with 13 datasets for the 5-year interval and 11 covering lifetime migration. Table 7.7 summarises the key features, linking trends apparent from the two migration measures. Turning first to the 5-year interval, it is clear that in most countries the trend was one of a decline in migration intensity. The most striking exception is China, where policy shifts coupled with rapid economic growth saw a tripling of inter-provincial migration, from 11 to 31 million from the 1980s and the 1990s.

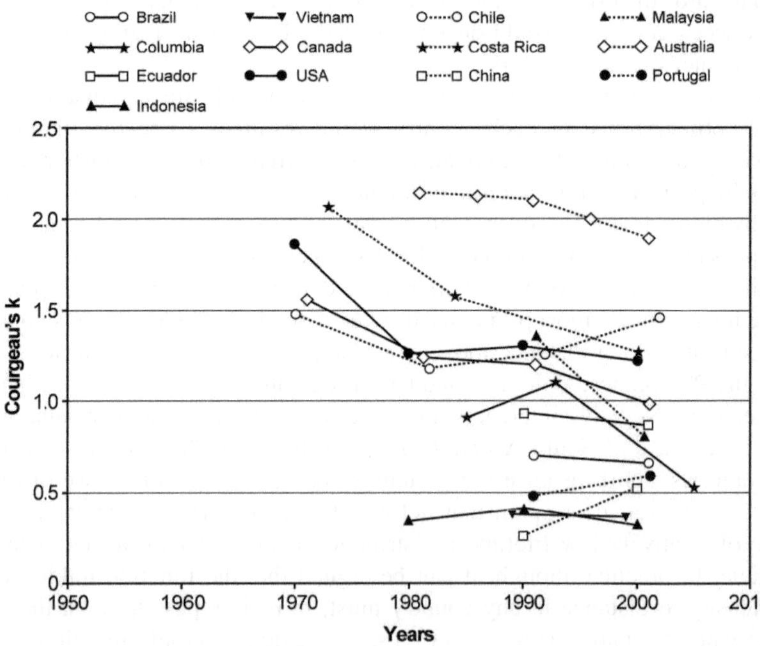

Fig. 7.3 Trends in Courgeau's *k* for 5-year migration intensities, selected countries (Source: Modified after Bell & Muhidin, 2009)

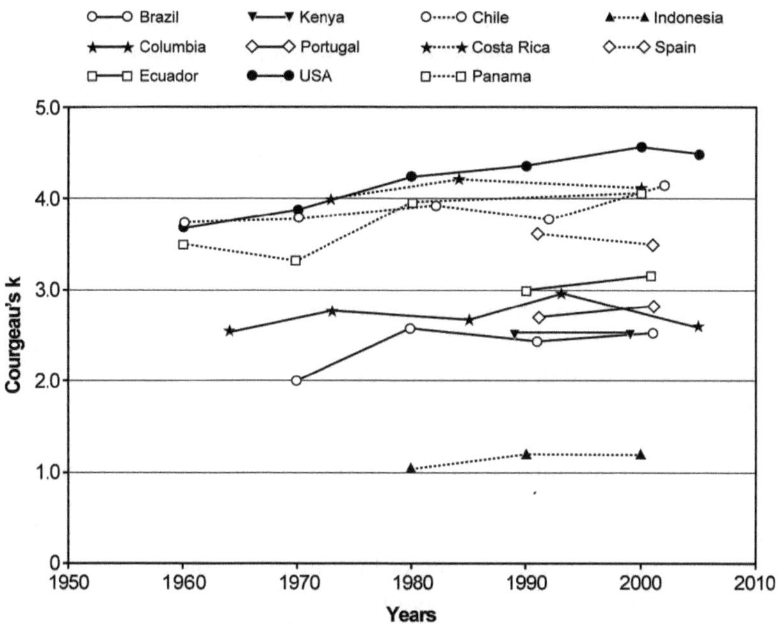

Fig. 7.4 Trends in Courgeau's *k* for lifetime migration intensities, selected countries (Source: Modified after Bell & Muhidin, 2009)

Table 7.7 Trends in migration intensity, selected countries

		Trends in	
Region	Country	Five-year intensity	Lifetime intensity
Africa			
	Kenya		Rise
Asia			
	China	Rise	
	Indonesia	Fluctuate	Rise
	Malaysia	Fall	
	Vietnam	Fall	
Latin America			
	Brazil	Fall	Rise
	Chile	Fluctuate	Rise
	Colombia	Fluctuate	Fluctuate
	Costa Rica	Fall	Fluctuate
	Ecuador	Fall	Rise
	Panama		Rise
Developed countries			
	Australia	Fall	
	Canada	Fall	
	Portugal	Rise	Rise
	Spain		Fall
	USA	Fall	Rise

Source: Modified after Bell and Muhidin (2009).

Portugal registered a more modest rise, and Chile likewise experienced rising migration intensities after 1980. In contrast the trend was relatively flat in Brazil, Ecuador and Indonesia, while the USA, Canada and Australia, together with Malaysia and Costa Rica recorded pronounced decline.

The graph for lifetime migration presents a strikingly different picture, characterised not by declines, but by increases in migration intensity for a majority of countries. Eight of the 11 countries in our sample recorded rising lifetime migration over the period, whereas this was the case for just two of 13 on the 5-year measure. Indonesia, Brazil, Chile, Costa Rica, Ecuador and the USA, all of which recorded falls or fluctuating trends on the 5 year measure, displayed rising lifetime intensities. Only in Spain, Colombia and the USA (for the latest 2000–2005 interval), is there a discernible negative trend in the lifetime index.

Our interpretation of these disparate trajectories in the two measures is that contemporary trends in inter-regional migration are continuing to generate displacements in the pattern of human settlement throughout the world, as captured in the lifetime measure, but that these increases are occurring at a decreasing rate. Lifetime migration is a cumulative process which inherits the effects of high period mobility and upwards pressure from the ageing of high mobility cohorts. These effects are minimised, and a more realistic picture of trends is provided, by examining intensities measured over a succession of fixed intervals. Our results demonstrate that, when this is done, the global picture is one of declining, rather than rising, mobility. What is equally apparent is that there are no clear regional differences in the underlying trends in migration intensity. Of the countries considered here, only China displays a stark rise in underlying population mobility, and that is readily explicable by reference to policy settings and economic conditions (Fang et al., 2009).

Conclusions

As spatial differentials in fertility and mortality continue to erode, interest in population mobility has gained increasing prominence, both among academics and policy makers. The year 2009 stands as a particular milestone in the evolution of migration studies, with two of the major international development agencies, the United Nations Development Program and the World Bank, both dedicating their flagship annual reports to the topic of human mobility (United Nations, 2009; World Bank, 2009). While these documents bring to the fore much that is of interest, they also underscore the deficiencies in contemporary measurement of population movement.

Notwithstanding sustained efforts since the 1970s to place the study of migration on a rigorous scientific foundation (see e.g. Rees, 1977), a reliable statistical basis for cross-national comparisons of population mobility has remained well beyond reach. This can be traced partly to persistent differences among countries in data collection practices and procedures; harmonisation of data types and migration intervals has barely begun. It also reflects the sluggish progress towards an internationally accepted set of migration indicators, comparable to those employed in other fields of demography. Ultimately, however, it is differences in the spatial domain that have presented the most substantial obstacle to progress, not only in regard to variations in the shape and size of countries, and in their patterns of human settlement, but in the statistical geography employed for data collection.

Courgeau's (1973) *k* represents an elegant and seductive approach to this problem, but has received surprisingly little attention in previous research. Its particular merit lies in making a virtue of the widespread differences in zonal systems that occur between nations. The index value *k* essentially corresponds to the slope of a regression line connecting migration intensities observed at different spatial scales, the latter captured as the natural logarithm of the square of the number of regions at which the intensity is observed. In theory, values of *k* therefore differentiate nations according to their levels of mobility, with *k* behaving according to a ratio scale: a value of $2k$ implies a migration intensity twice that of *k*.

We examined Courgeau's hypothesis using census data on 5-year and lifetime migration for a sample of 27 countries – the first comprehensive study of this measure. The analyses provide qualified support for the approach elaborated by Courgeau. The regression lines for individual countries form a fan like structure emanating broadly from the graph origin, and for those countries with more than three or more data points, the association appears close to linear. Moreover, when the lines of regression are forced to pass though the origin (for countries with two data points or more), the fit, as indicated by the coefficient of determination, was above 80% in 28 of 33 cases (combined 5-year and lifetime), and above 90% for 22.

The results show that *k* is effective as a measure discriminating between countries according to migration intensity. While there is no clearly delineated sequence of internal migration intensity by world region, some clustering is apparent. The developed countries all display high mobility, especially those in the 'new world', but Latin America too emerges as a region of high migration intensities. Asian countries, on the other hand, generally display the lowest levels of mobility, while the small number of African countries in our sample lie between the Asian and Latin American clusters. Similar patterns appear whether migration is measured over 5 years, or as lifetime intensity, but these divisions are most clearly apparent in the lifetime data. However, this broad sequencing disguises considerable overlap between regional clusters. Thus, within Latin America, there are stark contrasts between the high mobility evident in Chile, Costa Rica and Panama on the one hand, and the more subdued intensities apparent in Brazil, Argentina, Ecuador and Colombia on the other. Malaysia, with a *k* value double that of its fellow countries, is a notable exception to the relatively low intensities apparent elsewhere in Asia, and in Africa, high mobility in South Africa stands in marked contrast to the relatively low intensities registered in Ghana, Kenya and Uganda.

We assessed trends in internal migration by comparing *k* indices from successive censuses. Our results run counter to the conventional wisdom of a sustained rise in global population mobility over the last several decades. Measured using lifetime intensities, the trend is certainly upwards, with a majority of countries registering increases over time in the proportion of their populations living outside their region of birth. However, 5-year intensities overwhelmingly registered declines in the *k* index, with only China and Portugal showing a contrary trend. This discrepancy in the two measures is most easily reconciled as a progressively diminishing addition to the cumulative indicator (lifetime migration) from contemporary migration processes (the 5 year figure). Lifetime moves reveal the net effects of internal migration but have a high degree of intrinsic momentum. Moves measured over a fixed 5-year interval offer a more sensitive gauge of contemporary trajectories.

As a mechanism for comparative analysis of migration intensity, Courgeau's k has several valuable features. Foremost among these is that it enables contrasts to be drawn between countries with quite different zonal geographies. When used for trend analysis it automatically corrects for some aspects of boundary changes. It is simple to compute and can be implemented with only a small number of observations: indeed, it even provides a graphical space in which to situate countries for which only a single observation is available. However, several limitations are also apparent. Most critical, perhaps, is that the index value k has no intrinsic or plain language meaning. Thus, k is not directly interpretable as a demographic indicator in the same form as the total fertility rate, life expectancy or migration intensity. A second issue concerns the use of a common distance-decay parameter. While the power function adopted in this chapter provides a good general fit, none of the regression lines pass directly through the point of origin, the intercepts vary widely from positive to negative and there are numerous crossovers between regression lines which complicate interpretation.

Fitting regression equations to just two or three observations is inevitably a hazardous business and one obvious avenue for refinement is to introduce additional data points corresponding to migration intensities at alternative levels of spatial aggregation. While our analysis has simply mirrored the standard statistical geography found in individual countries, it is technically feasible to generate intensity observations at a range of spatial scales if a suitably disaggregated flow matrix is available. Calculations made using alternative aggregations of individual zones would also help address biases arising from the modifiable areal unit problem. Another possibility is to experiment with alternative formulations of the distance-decay function implicit in Courgeau's equation. Since countries differ widely in physical and human geography, it would be surprising indeed if they shared an identical intensity profile. One option would be to calculate country-specific distance-decay parameters using conventional spatial interaction models. Another would be to introduce measures of size, shape and settlement pattern as discrete explanatory variables. On any assessment, Courgeau's k seems worthy of further scrutiny in the pursuit of more rigorous understanding as to how, and why, levels of mobility differ between countries.

Acknowledgments This chapter draws on work undertaken for the United Nations Development Programme 2009 Human Development Report *Overcoming Barriers: Human Mobility and Development* (see http://hdr.undp.org/en/). We are grateful for permission to utilise here some of the material assembled for that project. The majority of the data utilised here are drawn from the Integrated Public Use Microdata Series (IPUMS) maintained by the Minnesota Population Center at the University of Minnesota: we are grateful to the IPUMS team for facilitating access to these data. We also thank colleagues at the United Nations Development Programme for facilitating access to data for China and India.

References

Australian Bureau of Statistics. (1995). *1991 census of population and housing, data quality – Undercount.* Canberra: Australian Bureau of Statistics.

Bell, M. (2003, May 1–3). *Comparing internal migration between countries: Measures, data sources and results*. Paper presented to the annual meeting of the Population Association of America, Minneapolis, MN.

Bell, M. (2005, July 18–23). *Towards rigorous cross-national comparison of internal migration: Who collects what?* Paper presented to the 25th conference of the International Union for the Scientific Study of Population, Tours, France.

Bell, M., Blake, M., Boyle, P., Duke-Williams, O., Rees, P., Stillwell, J., & Hugo, G. (2002). Cross-national comparison of internal migration: Issues and measures. *Journal of the Royal Statistical Society A, 165*(3), 435–464.

Bell, M., & Rees, P. (2006). Comparing migration in Britain and Australia: Harmonisation through use of age-time plans. *Environment and Planning A, 38*(5), 959–988.

Bell, M., & Muhidin, S. (2009). Cross-national comparisons of internal migration, *Human Development Research Paper 2009/30*. New York: United Nations.

Blake, M., Bell, M., & Rees, P. (2000). Creating a temporally consistent spatial framework for the analysis of interregional migration in Australia. *International Journal of Population Geography, 6*, 155–174.

Bogue, D. J., Ligel, G., & Kozloski, M. (2009). *Immigration, internal migration and local mobility in the US*. Cheltenham: Edward Elgar.

Boyle, P. J., & Feng, Z. (2002). A method for integrating the 1981 and 1991 GB census interaction data. *Computers, Environment and Urban Systems, 26*, 241–256.

Comisión Económica para América Latina y el Caribe (CEPAL). (2007). *Panorama Social de América Latina 2007*, LC/G.2351-P/E, Santiago do Chile.

Champion, A. G. (1989). *Counterurbanisation, the changing pace and nature of Population deconcentration.* London: Edward Arnold.

Courgeau, D. (1973a). Migrations et découpages du territoire. *Population, 28*, 511–537.

Courgeau, D. (1973b). Migrants and migrations. *Population, 28*, 95–128.

Fang, C., Du, Y., & Meiwan, W. (2009). Migration and labor mobility in China. *Human Development Research Paper 2009/09.* New York: United Nations.

Fielding, A. J. (1982). Counter-urbanisation in Western Europe. *Progress in Planning, 17*, 1–52.

Kitsul, P., & Philipov, D. (1981). The one year/five year migration problem. In A. Rogers (Ed.), *Advances in multiregional demography* (pp. 1–34). Laxenburg: International Institute for Applied Systems Analysis.

Langevin, B., Begeot, F., & Pearce, D. (1992).Censuses in the European community. *Population Trends, 68*, 33–36.

Law, G. (1999). *Administrative subdivisions of countries.* North Carolina: McFarland, Jefferson.

Long, J. F., & Boertlein, C. G. (1990). Comparing migration measures having different intervals. *Current Population Reports, Series P-23* (pp. 1–11). .Washington DC: US Bureau of the Census.,

Long, L. H. (1991). Residential mobility differences among developed countries. *International Regional Science Review, 14*,133–147.

Long, L. H. (1992). Changing residence, comparative perspectives on its relationship to age, sex and marital status. *Population Studies, 46*, 141–158.

Nam, C. B., Serow, W., & Sly, D. (1990). *International handbook on internal migration.*Westport, CT: Greenwood.

Newbold, K. B., & Bell, M. (2001). Return and onwards migration in Canada and Australia, evidence from fixed interval data. *International Migration Review, 35*, 1157–1184.

Rees, P. H. (1977). The measurement of migration from census data and other sources. *Environment and Planning A, 9*, 247–272.

Rees, P., Bell, M., Duke-Williams, O., & Blake, M. (2000). Problems and solutions in the measurement of migration intensities: Australia and Britain compared. *Population Studies, 54*(2), 207–222.

Rees, P., & Kupiszewski, M. (1996). *Internal migration and regional population dynamics: What data are available in the Council of Europe member states?* (Working Paper 96/1). Leeds: School of Geography, University of Leeds.

Rees, P., & Kupiszewski, M. (1999a). Internal migration: What data are available in Europe? *Journal of Official Statistics, 15*(4), 551–586.

Rees, P., & Kupiszewski, M. (1999b). *Internal migration and regional population dynamics in Europe: A synthesis.* Strasbourg: Council of Europe Publishing.

Rogers, A., Raquillet, R., & Castro, L. J. (1978). Model migration schedules and their applications. *Environment and Planning A, 10*(5), 475–502.

Rogers, A., & Castro, L. J. (1981). *Model migration schedules. Research Report RR-81-30.* Laxenburg, Austria: International Institute for Applied Systems Analysis.

Rogerson, P. A. (1990). Migration analysis using data with time intervals of differing widths. *Papers of the Regional Science Association, 68*, 97–106.

Stillwell, J., Bell, M., Blake, M., Duke-Williams, O., & Rees, P. (2000).A comparison of net migration flows and migration effectiveness in Australia and Britain: Part 1, total migration patterns. *Journal of Population Research, 17*(1), 17–38.

United Nations. (2000). *World population monitoring 1999: Population growth, structure and distribution.* Population Division, ST/ESA/SER A/183, Department of Economic and Social Affairs. New York: United Nations.

United Nations. (2008). *Principles and recommendations for population and housing censuses, revision 2.* Statistical papers Series M No. 67/Rev.2 ST/ESA/STAT/SER.M/67/Rev.2. New York: Department of Economic and Social Affairs.

United Nations. (2009). *Overcoming barriers: Human mobility and development.* 2009 Human Development Report, United Nations Development Programme, New York.

Vignoli, J. R. (2004). Migración interna en América Latina y el Caribe: Estudio regional del período 1980–2000, LC/L 2059-P, *Serie Población y Desarrollo, No 50*, Centro Latinoamericano y Caribeño de Demografía (CELADE, División de Población de la CEPAL), Santiago de Chile.

World Bank. (2009). *World development report 2009: Reshaping Economic Geography, Report Number 43738,* Washington, DC: The World Bank.

Wrigley, N., Holt, T., Steel, D., & Tranmer, M. (1996). Analysing, modelling, and resolving the ecological fallacy. In P. Longley, & M. Batty (Eds.), *Spatial analysis, modelling in a GIS Environment* (pp. 23–40).Cambridge: GeoInformation International.

Xu-Doeve, W. L. J. (2006, June 21–24). *The measurement of international and internal migration in the 2010 global round of population censuses: Twelve key recommendations on questions, concepts and procedures.* Paper presented at the European Population conference of the European Association for Population Studies (EAPS), Liverpool.

Chapter 8
Increasing Longevity and the Economic Value of Healthy Ageing and Longer Working

Les Mayhew

Introduction

One of great success stories in the United Kingdom (UK) is that people are living longer. Life expectancy at birth is now almost 80 years, having advanced 11 years since 1950 thanks to improvements in occupational health, health care, fewer accidents and higher standards of living. This success in turn presents a huge opportunity for individuals if extra years are spent in prosperity and good health. As a recent Foresight report (Foresight, 2008) has demonstrated, tapping into the experience and skills of older adults can also benefit employers, public services and voluntary and civic organisations. Indeed, realising the full potential of older citizens of the UK will be central to the Government's response to changing economic circumstances and the drive to build a strong, fair economy for the twenty-first century. However, the challenges posed by an ageing society do not rest solely with older citizens. Referring to the high levels of economic inactivity, the Black report noted: 'the sheer scale of people on incapacity benefits represents an historical failure of healthcare and employment support to address the needs of the working age population in Britain' (Black, 2008, p. 13). In other words, the health of the working age population is needed to sustain the economy of the whole population and this must not be overlooked as it is part of the solution.

Studies on population ageing usually take one of three forms: analysis of macroeconomic problems relating to the decline in the workforce; analysis of social security systems; and labour market studies, many focusing on older workers (Mackellar, Ermoliieva, Horlacher, & Mayhew, 2004). In this chapter, we tackle similar issues but do so from a health perspective. The hypothesis is that as a population ages, health tends to deteriorate and this creates problems for the economy, whilst good health becomes a scarcer commodity, leading to increased healthcare costs, high social security benefits and lost economic output. Hence it is important to quantify these effects and compare them with strategies that maintain or improve health.

L. Mayhew (✉)
Cass Business School, Faculty of Actuarial Science and Insurance, City University, London, UK
e-mail: lesmayhew@googlemail.com

J. Stillwell, M. Clarke (eds.), *Population Dynamics and Projection Methods*, Understanding Population Trends and Processes 4, DOI 10.1007/978-90-481-8930-4_8,

The extent to which this will apply to the UK is additionally important for the following reasons. The UK is going though a very rapid period of population growth from around 55.9 million in 1980 to a projected 71 million by 2030. Over the same period, the structure of the population will alter significantly with the population aged 65+ doubling from 8 million in 1980 to 16 million by 2030. The ratio of people aged 20–64 to 65+ was 3.7 in 2008, much the same as in 1980, but from now onwards it is due to go into rapid decline reaching 2.5 by 2030 as the population enters an era of unprecedented ageing. In addition to this, the UK total support ratio (TSR),[1] which is often associated with the rapid economic ascendancy of countries like Japan, peaked in 2007. The rise up to 2007 is the result of falling fertility and the decline after 2007 is caused by the progressive retirement of baby boomers.

An explanation for such demographic transitions is as follows. Initially fertility is high and accompanied by low child support ratios (i.e. a small ratio of adults to children) as occurred after the war. With demographic transition, the proportion of working age population to total population increases, favouring labour supply and improved savings rates. The flipside occurs when fertility falls and the population ages and so that a low child support ratio gradually transforms into a low old age support ratio. In between, the TSR peaks and then declines, albeit along a slow glide path and so the demographic cycle is complete. It follows that the favourable labour market conditions created by the rise in the TSR needs to be replaced by other factors, such as full employment and improved education and health levels in order to maintain economic growth (Fang & Wang, 2006).

The UK transition is on a smaller scale and lagging behind Japan by about 15 years but the potential consequences are no different. One obvious consequence is pressure to increase pension age. To maintain the old age support ratio at today's level of 3.7, UK state pension age would need to increase to 65.5 years in 2020, 66.5 years in 2025 and 67.5 years by 2030 (Mayhew, 2009). Built into this reckoning are assumptions that people will work longer and be more productive, because the alternative is that taxes will have to rise or the working age population would need to be bolstered through higher levels of migration rather than increased fertility. The danger to economic growth is that historical trends in wage productivity will slow or stagnate for the following reason. Analysis shows that earnings peak in the 35–49 age group, and that in 50+ age groups earnings decline. The population in the 35–49 age group has recently peaked and therefore its contribution to earnings growth and hence productivity will decline too.

It is debateable whether the trend will be as mechanistic as this in the future, as the over 50s may have skills that are valued compared to today's over 50s. Nevertheless, a demographic shift could render the possibility of stagnant wages a real possibility. Growth will be determined by whether productivity of older workers is lower or whether older workers provide an adequate replacement for

[1] The total support ratio or dependency ratio, which is the ratio for the number of people aged 20–64 to the population < 20 plus population 65+, peaked in 2007 after a steady rise for over 27 years.

younger workers. Labour competes in a global economy and firms can source their production in countries with a plentiful supply of low cost labour. Much will depend on the nature of the work and on skills, e.g. 'old' economy workers are 'burnt out' at an earlier age than knowledge workers in the 'new' economy (Blake & Mayhew, 2006). However, there are dangers lurking behind these simple assumptions which depend crucially on extra years spent in work being healthy years as well as on continued economic growth.

A pessimistic scenario is that an ageing population will simply increase the stock of unhealthy people resulting in lower productivity and more people under care. Poor health is not confined to older age groups and a significant number of working age adults are economically inactive due to long-term sickness and disability. Based on measures reported in this chapter, today there are 5.1 healthy adults aged 20+ for every unhealthy adult; by 2025 this could fall to 4.3 to 1 due to ageing. To put this in perspective, there would either need to be an increase of 8 million in the number of healthy adults in order to maintain the current balance, or healthy life expectancy at age 20 would need to increase by about 3.5 years given the expected increases in life expectancy. Were we only to include healthy people below state pension age in this calculation, we would find that increasing pension age would not be able to restore the level to 2007 levels since we would soon run out of healthy people! On the other hand, this might be possible if there were only reasonable health improvements at every age, in which case state pension age could be held at 68.

Of the £250 billion the UK spends each year on healthcare, social security benefits and social care, about £30 billion is spent on benefits for the long-term sick and disabled, and £20 billion on social care. The share spent on healthcare for the long-term sick and disabled is harder to calculate but is somewhere in the region of £40 billion. These figures suggest that average annual public expenditure on the estimated 7.3 million long-term sick and disabled adults is around £10,000–£13,000 per person per year depending on one's assumptions. Social security benefits must be paid for through taxes or out of pocket expenditure. Benefits paid in kind such as caring activities are generally paid for by foregone wages and economic output depending on the age of the carer. If this already sizeable problem could be tackled by health improvements, it may be possible to redress some of the balance in these support ratios, at least in part. This requires both a more detailed understanding of the demographic trends in health, coupled with work and also some means of quantifying the scale that different health improvements and interventions could make to the equation.

A difficulty is that there are numerous measures and definitions of health and so what we mean by ill-health or disability is, to a considerable extent, arbitrary. We use terms such as morbidity, physical disability, self-reported health or benefit eligibility interchangeably. One measure of health corresponds to a person that would qualify for one or more of the current sick and disability benefits depending on severity, but other measures and definitions must also employed (e.g. to enable international comparison, or with reference to morbidity rather than physical disability). Based on benefit eligibility, there are 2.8 million adults that fall into this definition aged

20–64 and 4.5 million aged 65+. Based on the age profiles of current claimants, the equivalent figures in 2025 will be 3.1 and 6.6 million.

With the sharp downturn in the old age support ratio and the rapidly expanding number of older people, the evidence suggests that we are on the threshold of a new era in UK history that is set to continue for the foreseeable future. The next 15–20 years provide the window of opportunity for putting in place the necessary policies and systems to support them. This chapter is concerned with explaining these trends in some detail to provide:

- estimates of the potential downside of getting outcomes wrong or doing nothing based on current trajectories; and
- estimates of the potential economic upside of getting outcomes in later life right, with a focus on better health and greater participation (healthy, active ageing).

The chapter builds on and analyses three key quantities: life expectancy (LE), healthy life expectancy (HLE) and working life expectancy (WLE). Various hypotheses follow from changes in these quantities. For example, a 'downside' scenario could be further rises in LE but no corresponding increases in HLE or WLE. This could significantly increase the health burden with corresponding falls in living standards and a rise in population due to increased demand for migrant labour. An 'upside' hypothesis, which we call the 'active ageing' scenario, would result in a narrowing in the gap in LE and HLE and increases in WLE. This would result in improved living standards and alleviate migration pressures.

Many important questions relate to these measures and the differences between them. For example, what does closing the gap between LE and HLE by 2 years and extending labour participation rates beyond state pension age do for Government expenditure/revenue and Gross Domestic Product (GDP) growth (the so-called healthy 'active ageing scenario')? Alternatively what does widening the LE and HLE gap and holding labour participation rates constant do for Government expenditure/revenue and GDP growth (the unhealthy, 'passive ageing scenario')? The answers to such questions for the UK are crucial if the aim is to continue to increase GDP in a globally competitive world, but also to maintain or increase living standards (since the two are not necessarily the same thing).

By the arguments put forward in this chapter, these objectives can be achieved by different means, but not all equally desirable: for example, (i) by simply allowing the population to grow unrestrained; or (ii) by pursuing a more orderly approach in which the full potential of the population is realised through better health and economic engagement. In the real world, the economy could be easily overwhelmed by other economic factors unrelated to demography, but ignoring demography would leave too much to chance. A key advantage of the simplistic approach taken is that it is possible to isolate the variables that support the general argument and draw simple conclusions. Based on the model, we show how changes to LE, HLE, WLE could affect various areas of Government expenditure, taxes and GDP and we use the model to consider the changes needed to put the UK economy on an 'active ageing path'.

Life, Healthy Life and Working Life Expectancy

There is no single source of data on LE or HLE that serves all purposes. Life expectancy is the average number of years of life remaining at a given age. Healthy life expectancy is a generic term for any of a number of summary measures which use explicit weights to combine HLEs for a set of discrete health states into a single indicator of the expectation of equivalent years of good health at a given age.[2]

To make our arguments, we begin by comparing the UK on a range of published measures used by the World Health Organisation (WHO), the Office of National Statistics (ONS) and the Human Mortality Database. Clearly our analysis needs to be more flexible and detailed than these data can provide. For example, we are interested in measuring LE and HLE at different ages and not just at birth, and we would like to be able to disaggregate these quantities into the experiences of different socio-economic groups and lifestyles. We combine these measures to analyse the general effects of an ageing population on the UK economy of trends and changes in their relative values over time. In order to produce the insights we need, we use a simple uncluttered framework combining key variables rather than attempt to model the whole economy in detail. Hence LE is important because it affects how people plan their lives and spend their time; for example, whether to invest for longer in education, to save or to consume. A high HLE creates the necessary conditions for any economic activity to be undertaken and influences the decision to remain economically active for longer and thus increase WLE. Higher WLE is associated with economic growth, investment in research and development and improved quality of life. LE is greater than HLE which in turn is greater than WLE. The first must be true and the second is generally true. The difference between LE and HLE can be interpreted as the number of years spent in ill health and disability (usually, but not exclusively at the end of life). The difference between HLE and WLE can be regarded as the healthy years spent in economic inactivity (broadly leisure, in retirement, caring, house keeping and education). From a societal point of view, WLE can be thought of as being constrained by three factors: years in education, years spent in caring activities on behalf of others (mainly children and older people), and law (e.g. minimum or maximum ages in the work force).

---

[2]Disability-free life expectancy measures disability by looking at reported limitations in day to day activities such as work. The World Health Organization (WHO) defines a quantity known as HALE (health adjusted life expectancy) as the average number of years that a person can expect to live in full health. HALE is calculated by subtracting from the life expectancy the average number of years in ill-health weighted for severity of the health problem. The first example of 'health expectancy' was published in a report of the US Department of Health Education and Welfare (Sullivan, 1971).

Healthy individuals are generally more productive than unhealthy ones and are more flexible in terms of the work they do and thus finding employment. A low HLE may conflict with a policy of 'active ageing' if it results in early withdrawal from economic activity ahead of pension age, and if HLE is less than pension age, then a person may require financial support through the benefits system (i.e. disability benefits). The lesser the gap between LE and HLE, the lower is the prevalence of disability and ill health in society, whereas the greater the gap, the more people will be dependent on health and social care and the more healthy people will be diverted into caring activities. Closure of the gap is also termed the 'compression of morbidity' and means that illness is compressed into a smaller number of years over the life cycle (Fries, 1980).

A country which scores badly on any of these indicators in which the gap between any of them is excessively large, will therefore tend to suffer economically through low growth, productivity and potentially higher taxes. There is evidence of a strong impact of increased LE on economic performance, namely that increases in GDP per capita are associated with increases in LE. The finding that LE increases with income, albeit at a diminishing rate, seems to hold regardless of whether studied at the global, national, community or individual level; but it also holds across demographic groups and in different economic contexts. This seems highly intuitive as life extension occurs at a diminishing rate and usually each extension costs more than the previous one. One question is that if health care resources were allocated such that those benefiting experienced the largest improvement in LE per unit of resource spent, would LE increase at a faster rate? It seems plausible that it would since potential gains in LE for those at the top of the LE tree are likely to be smaller than for those at the bottom.

It has been suggested that income inequalities could also have a direct impact on individual health and therefore LE (Kawachi et al., 1997). Hence, the observed negative correlation between LE and income inequality could be the result of diminishing returns or an actual causal effect, and in this regard several mechanisms have been proposed. It is argued, for example, that societies with sharper inequalities tend also to suffer from a lower level of social capital and mutual trust, which in turn might be detrimental to health. Due to the lack of social cohesion, individuals are exposed to higher crime or accident rates, which have a direct impact on health. Finally, unequal communities tend to be more polarized and might, as a result, provide unequal access to public services (Araujo et al., 2008; Krugman, 1996; Zhao, 2006).

Karlsson, in an unpublished paper,[3] has studied the relative effects that a difference in absolute income would make compared with a reduction in income inequality. He found, for example, that a $1,000 increase in the GDP per capita in the UK (at $29,462 in 2004, the latest year in the international dataset used) would have bought 0.11 additional life years. This figure could be compared with the estimated effect of a similar increase in India (GDP per capita $3,213) where

[3]Life expectancy, GDP and inequality (personal communication).

the same increase in GDP would buy an additional life year. Similarly, reducing the UK income inequalities as measured by the Gini coefficient[4] (currently 0.32) to the lowest level recorded in 2004 (Sweden, 0.23) would increase LE by 0.16 years. He notes that eliminating inequalities altogether would increase LE by another 0.41 years provided the assumption of a linear effect for all levels of inequalities is correct. Interestingly, he also points out that even though the inequality effect might appear to be more important than the absolute income effect, real increases in GDP of this magnitude occur in a much shorter space of time than it would take to reduce inequalities to achieve a similar result. He goes on to show that GDP per capita also produces similar effects on HLE, so implying that identical arguments will apply to HLE as apply to LE. This is important since it has been argued that findings of an association between inequality and health could be attributable to 'reverse causality', i.e. policies which improve health or educational attainment amongst the poor are also likely to reduce income inequalities over time so that investment in either would confer important long term benefits through increases in social and human capital.

The Gap Between Life Expectancy and Healthy Life Expectancy

In international comparative terms, UK LE and HLE are up with other developed countries as one might expect but are not in the vanguard. The World Health Organization (WHO), for example, shows that the gap between LE and HLE for different life expectancies at birth is more or less constant regardless of LE and that the average years spent in poor health is equivalent to 10 years of life at birth. In developing countries with a low LE a far greater proportion of life is therefore spent in ill health and disability than in more developed countries. For developing countries, low LE and ill health tend to be related to infectious diseases and in developed countries to the chronic diseases of older age.

Based on the concept of HALE (Health Adjusted Life Expectancy ~ see footnote 2), WHO shows that the UK is ahead of the US in HLE and LE but behind Japan which has both the highest LE and HLE in the world and also the smallest gap between LE and HLE. UK LE is 79 years, HLE 71 years (gap 8 years), LE in Japan is 82 years, HLE 75 years (gap 7 years) and for the US LE is 78 years, and HLE 69 years (gap 9 years). These data are based on a 2003 snapshot and do not therefore show how either LE or HLE are changing over time.

The Office for National Statistics (ONS) publishes statistics on HLE for Great Britain, which it defines as years of expected life in either good or fairly good health

[4]The Gini coefficient is a measure of statistical dispersion ranging in value from 0 to 1 and is used as a measure of income or wealth inequality. A value of zero corresponds to perfect equality (everyone having exactly the same income) and 1 to perfect inequality (where one person has all the income, while everyone else has zero income). Worldwide, Gini coefficients range from approximately 0.232 in Denmark to 0.707 in Namibia. More advanced economies tend to have a Gini coefficient of between 0.25 and 0.50.

(based on general health) or free from long standing illness.[5] ONS data suggests that whilst both LE and HLE are increasing, the gap between them is widening. Trend analysis of ONS data since 1981 shows that:

- LE at birth will be 83.2 years in 2025 as compared with 79.1 years in 2005 (the latest year for which data are available), an increase of 4.1 years;
- HLE will be 71.7 years at birth in 2025 as compared with 69.3 years in 2005 an increase of 2.4 years; and
- by 2025, the gap between LE and HLE will be 11.48 years compared with 9.75 years in 2005, equivalent to an average change of 28.8 days per annum.

If correct, the above in turn implies 86% of life was spent in good or fairly good health in 2005 as compared with a similar percentage that will be spent in 2025. The difference is that the period of life spent in disability in 2025 will be greater with consequent impacts, mainly on health care and older people's services.

Disability-Free Life Expectancy and Disease-Free Life Expectancy

The ONS uses other definitions of health based on being free of disability which tend to suggest more years are spent in disability though not necessarily in poor health. Using the Health Survey for England (HSE), Rasulo, Mayhew, and Rickayzen (2008) compared two variants; one is disease-free LE and the other disability-free LE for the population aged 16 and over. The HSE includes questions on the occurrence of long-term and limiting long-term illness, and on the occurrence of conditions that require medicine to be taken regularly. Respondents with a long-term illness could list up to six illnesses while, for each prescribed medicine, the survey provided the corresponding disease under treatment. The questions on long-term illness and medicine were used to obtain the wider measure of morbidity, which was called 'life expectancy with disease'. The question on limiting illness, used for the computation of disability-free LE, was included for the first time in 1997 when individuals reporting a long-term condition were also asked whether this condition was limiting their daily activities. Reported diseases and disabilities by survey respondents were aggregated into categories. These reflected a combination of trauma, chronic and long-term conditions, as well as infectious diseases and acute episodes.

The key results are shown in Table 8.1 and indicate that LE is increasing for both males and females but the increase is larger for males, and that LE with disease has increased more for males than for females. It is particularly noteworthy that most of

[5]Two types of HLE are routinely calculated from national General Household Survey based on either of the following questions: 'Over the last 12 months would you say your health has been good, fairly good, or not good?' and LE free from limiting long-term illness based on: 'Do you have any long-standing illness, disability or infirmity?'. The method used by ONS to derive health expectancy is known as the Sullivan Method (Sullivan, 1971; Breakwell & Bajekal, 2005).

Table 8.1 Life expectancy, disease-free LE and disability-free LE at 16 in 1998 and 2008 in England

	Males			Females		
Category of expectancy	1998	2004	Difference (years)	1998	2004	Difference (years)
Life expectancy	59.7	61.5	1.8	64.5	65.7	1.2
Disease free life expectancy	29.4	29.6	0.2	28.3	28.5	0.2
Disability free life expectancy	44.4	46.1	1.7	46.1	47	0.9
Years spent with disease	30.3	31.9	1.6	36.2	37.2	1.0
Tears spent with disability	15.3	15.4	0.1	18.4	18.7	0.3

the additional years are being spent with non-limiting diseases, which is slightly less of the additional years being spent with disability and that most additional years are being spent with co-morbidity as opposed to a single disease. For example, the co-morbidity category, 'cardiovascular, respiratory or other chronic diseases, and other acute diseases', was a significant cause for increasing both disabled and disease life expectancies.

For some purposes, such as estimating the demand for long-term care a more appropriate measure is disability-free LE at age 65. International comparison shows that the UK does less well than competitors in either Japan or Germany. According the ONS, disability-free LE at age 65 is 10 years, which is an improvement over recent years. However, this is below levels in Japan, Germany, Netherlands or Switzerland which all achieve over 12 years (OECD).[6]

Impact of Poor Health and Increased Longevity on Taxes and the Economy – A Simplified Model

The significance of these findings is illustrated with the aid of a highly simplified model of the economy. The description that follows is intended as a framework for investigating the effects on taxes, living standards and GDP as a result of changes in key quantities, such as LE, HLE, WLE and wage productivity. The aim is to show how changes in one variable affect the economic variables of interest and scenarios that cause living standards and/or GDP or taxes to rise or fall. Consider a situation in society in which the working age population crudely divides into one of two groups consisting of either healthy or unhealthy people. The unhealthy group

[6]Disability-free LE is defined as the average number of years an individual is expected to live free of disability if current patterns of mortality and disability continue to apply. Disability definitions and measurements are only partly harmonised across countries.

do not work and receive financial support from the state or they are retired and receive a pension plus additional financial support for their disability. The healthy group either work or are economically inactive and if they are retired they receive a pension. The economically inactive population are in caring roles, unpaid work, and full-time education or simply in leisure.

Without loss of generality we focus on the population aged 20+ and define the following quantities from age 20: expected total life, expected working life (alive and under state pension age), expected retired life (alive and over state pension age), expected healthy working life, and expected healthy retired life. In this framework HLE is simply expected healthy retired life plus expected healthy working life.

Variables are introduced for average earnings, pension value, and benefits rates so that we can derive values for taxes, GDP and GDP per capita dependent on the values of the quantities above for a population, in this case the UK. For simplicity we assume that total wages are a proxy for GDP (i.e. we ignore investment income, rents *et cetera*).[7] The questions we wish to ask relate to the values all these quantities might take. This enables us to evaluate the importance of different variables in the model such as health and LE and to relate them to financial quantities such as wage productivity and benefit rates. We define the following quantities (all values calculated at the same point in time):

e_l = expected total life remaining;
e_w = expected working life (alive and under state pension age);
e_r = expected retired life (alive and over state pension age);
e_{hw} = expected healthy working life;
e_{hr} = expected healthy retired life;
e_h = healthy life expectancy;

and we observe the following identities:

$$e_l = e_w + e_r \tag{8.1}$$

$$e_l = e_{hw} + (e_w - e_{hw}) + e_{hr} + (e_r - e_{hr}) \tag{8.2}$$

In words:
Expected life = expected healthy working life + expected unhealthy working life + expected healthy retired life + expected unhealthy retired life

Other quantities of interest are the proportion of sick and disabled in the stable population and the proportion of healthy people:

$$d = \frac{(e_l - e_h)}{e_l} \tag{8.3}$$

[7] A full blown demo-economic model would include the whole population, expenditure on defence, education, servicing debt *et cetera* and importantly non-wage GDP.

$$h = \frac{e_h}{e_l} = 1 - d \tag{8.4}$$

where $e_l - e_h$ equals the expected years in disability.

Assume that when people are in the status of 'ill or disabled' they cannot work. Further, assume that benefit payments received is the value of benefits and health care received. Define the following:

$a =$ participation rate (% of healthy lives of working age that work);
$w =$ average wage;
$p =$ pension;
$b_w =$ sickness and health benefits paid to people of working age; and
$b_r =$ sickness and health benefits paid to people of retired age in addition to pension

We can consider the average individual aged 20 and get the following results (assuming no inflation) for the lifetime wages earned and benefits received:

(i) total wage : $w_{sum} = e_{hw}aw$ (8.5)

(ii) total benefit received when working age: $b_{wsum} = (e_w - e_{hw})b_w$ (8.6)

(iii) total pension received when retired: $p_{sum} = e_r p$ (8.7)

(iv) total additional benefit received when retired and in ill-health:

$$b_{rsum} = (e_r - e_{hr})b_r \tag{8.8}$$

Assuming no investment return, then the tax rate t needed for the individual to be 'self supporting', i.e. they pay sufficient tax when working to pay for their benefits, is:

$$t = \frac{b_{wsum} + p_{sum} + b_{rsum}}{w_{sum}} \tag{8.9}$$

If we assume that the population is stable, i.e. stationary with constant births and deaths, then we can simply calculate aggregated values for the entire population by multiplying the above variables by the factor:

$$f = \frac{P_{20+}}{e_l} \tag{8.10}$$

where P_{20} is the population age 20+.

Total aggregated wage is then: $w_{sum} f$
Total benefit paid to population of working age is: $b_{wsum} f$
Total pension paid to population of retired age is: $p_{sum} f$
Total additional benefit paid to retired population who are ill is: $b_{rsum} f$

Then, assuming that benefits are paid on a PAYG (Pay As You Go) basis (i.e. no surplus fund is built up) then the tax rate, *t* is:

$$t = \frac{(b_{wsum} + p_{sum} + b_{rsum})f}{w_{sum}f} = \frac{b_{wsum} + p_{sum} + b_{rsum}}{w_{sum}} \tag{8.11}$$

i.e. the same as the individual rate.

For large periods, one or more of these values will be constant. For example, if we assume no changes to the working population, wages or benefits then both w_{sum} and b_{wsum} are constant. If we increase life expectancy in old age but keep the number of years spent in ill-health the same then p_{sum} changes but not b_{rsum}. The benefit of this model is that by isolating the constituent parts one can see the true effect of increasing only one of the variables. Assuming GDP can be represented by total wages then:

$$GDP = e_{hw}aw\frac{P_{20+}}{e_l} \tag{8.12}$$

and GDP per capita by:

$$g = \frac{GDP}{P_{20+}} = \frac{e_{hw}}{e_l}aw \tag{8.13}$$

This states that the GDP per capita is equal to the proportion of the population that is healthy and of working age multiplied by the percentage of this potential working population who actually work multiplied by the average wage. Therefore, GDP per capita increases if the:

- proportion of population that is classed as working age increases, i.e. if the state pension age is increased;
- proportion of population of working age that is healthy increases;
- proportion of healthy working age that work increases; and
- average wage increases (as this is the proxy of GDP).

We contrast situations at two points in time using numbers that approximately correspond with current experience and at a point in the future chosen arbitrarily to be 2025 for illustrative purposes. Figure 8.1 shows a structure of the simplified model of the economy based on this framework. Only non-wage GDP has been omitted from the calculations that follow (i.e. GDP generated from dividends and rents including overseas earnings). Typical questions to be asked of the model would be by how much would taxes need to increase if there were an increase in LE but no corresponding increase in HLE, *or* what would be the effect on GDP/capita (a broad measure for standard of living) of a health improvement with other variables remaining the same?

The scenarios are designed to cover a range of possible futures; they include a baseline corresponding to the present a continuation of present trends (the passive ageing scenario), health deterioration (worst case) and accelerated improvement (the

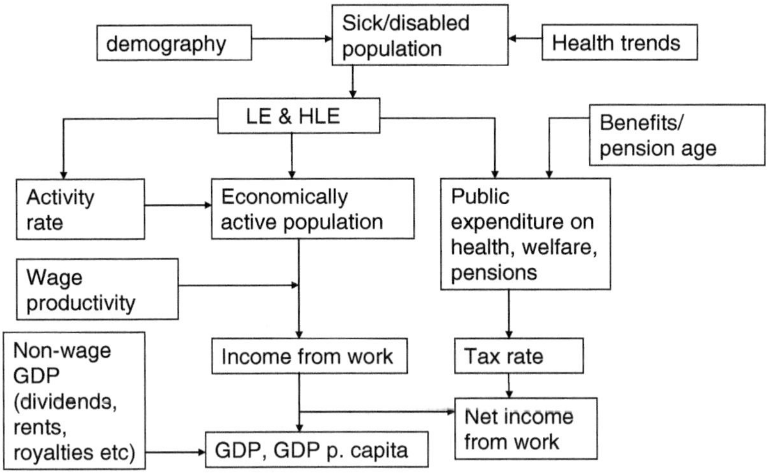

Fig. 8.1 Flow diagram showing the structure of simplified model of the economy

Table 8.2 Table of expectancies used in scenarios

Scenario	e_l	e_h	e_w	e_{er}	e_{hw}	e_{hr}	Years of life spent in disability	Years in retirement in ill health	Years of working age in ill health
Baseline	60	50	42	18	37	13	10	5	5
A	63	52	42	21	38	14	11	7	4
B	63	49	42	21	36	13	14	8	6
C	66	56	44	22	39	17	10	5	5

'active ageing' scenario). Specific values for the parameters used in each scenario are given in Table 8.2 and the economic quantities in Table 8.3. To give an example in 2007 the average wage was £23,000, having grown historically at a real rate of around 2% over annum over the long term.

The benefit rate is assumed to be £10,000 and corresponds with the lower end of earlier estimates based on current levels of benefit expenditure, health and social care, whereas the state pension is valued at £5,000 per annum. Assumed life expectancy, HLE, and WLE in the base period are fixed at 60, 50, and 42 years[8] respectively. This gives a tax rate of 29.4%[9] and after tax wage of £16.200, a wage-GDP of £408 billion based on the wage sum and a GDP/head of £9,077 (Table 8.3, row 1). Now consider the following three different scenarios based on a 2025 horizon which we compare to the baseline case:

[8]This is an estimate based on the Labour Force Survey (LFS) which shows that male WLE based on being classed as economically active is 39 years and females 37 years. If periods of unemployment are included the averages fall to 35 and 33 years respectively.

[9]The average tax rate is based solely on the costs of health and social care, social security benefits and state pension.

Table 8.3 Scenario results showing implied increases in taxes and changes in wage rates consequent on the assumptions given

Scenario	LE @ 20	HLE @ 20	Years in disability	Participation rate	Average wage (£000s)	Tax rate (%)	After tax wage (£000s)	GDP £bns	GDP/head
Baseline	60	50	10	0.64	23	29.4	16.2	408	9,077
A	63	52	11	0.65	27.5	25.5	20.5	555	10,786
B	63	49	14	0.63	19.2	45.3	10.5	355	6,910
C	66	56	10	0.67	32.8	30	23	669	13,005

A. *Passive ageing (current indicated trends)*: Health expectancy improves 3 years, life expectancy increases to 63 years. Health gains accrue in pre- and post-retirement age, years spent in disability widens to 11 years. Participation rate increases to 65% and wage productivity increases by 1% p.a.
B. *Health deterioration*: Health expectancy falls to 49 years because of fall in health in pre-retirement age but LE at 20 increases from 60 to 63 so that 14 years of life are spent in disability. Participation rates fall to 63%, and wages fall by 1% p.a.
C. *Active ageing*: Life expectancy at 20 increases to 66 years and HLE to 56 years maintaining baseline gap of 10 years in disability. Additional healthy years are spent in pre- and post-retirement; pension age increases by 2 years. Participation rates increase by 3% and wages, pension and benefit rates by 2% p.a.

Scenario C is described as the 'active ageing' scenario because it delivers longer life, better health and wages but also higher benefits for disabled people for very little change in tax rates as compared with baseline.[10] The worst case is scenario B in which life expectancy continues to increase but the gap between LE and HLE expands by 4 years and participation rates go down. Clearly there is a multitude of other possible scenarios. The implied change in tax rates and wages resulting from each scenario are given in Table 8.3 alongside GDP/capita and wage-GDP. It shows that any scenario that involves improvements in health relative to LE, increases in participation rates, or improvements in wage productivity delivers lower taxes and higher net wages, and greater GDP/capita etc.

In calculating wage-GDP[11] itself, the results are scaled by the size of the population. The UK's population age 20+ is due to increase from 44.5 to 51.4 million in the period and the whole population from 55.9 to 71.1 million. The 2025 population is used in each of the scenarios except for the baseline. Three illustrative cases may be contrasted:

1. As per scenario B with more years spent in disability. Wage-GDP falls and there is decline in participation rates. In other words, the wealth of the country and standards of living decline.
2. As per scenario A, the 'passive ageing scenario', with health improvement at older ages and modest wage and participation rate improvements. GDP and GDP per capita increase and tax rates would be reduced due to lower disability benefit payments.
3. As per scenario C, improved health, wage productivity and participation rates deliver much higher GDP and GDP/capita, but comparable benefit and pension

[10]For a more in-depth treatment of the concept of 'active ageing' see for example ICCR (2005).

[11]GDP is the main measure of national income. In economic theory, national income equals national expenditure which equals national product. Our simplified representation using a proxy based simply on wage income and so ignores other sources of income *et cetera*.

increases to wage increases result in a higher tax rate but also a higher after tax wage.

To give a further illustration, consider two cases, one with health improvements and one without. Without health improvements, no increase in economic activity and an assumed annual increase in wages of 2%:

- GDP per head would increase at an average of 1.5% per annum;
- public expenditure on health, welfare benefits and pensions would go up 92% over the period to 2025 due to an ageing and growing population; and
- average tax rates on earnings would increase 10% points.

If two more years of life were spent before pension age in good health, and two more years of health in remaining life, and there was a 3% increase in economic activity:

- GDP per head would increase at 1.9% p.a., almost the same as wages;
- public expenditure would go up by 68% instead of 92%; and
- average tax rates would increase by 1.9% points.

It must be emphasised that this simple model is not a predictive tool and is used only for indicative purposes and relies heavily on the assumption of a stationary population based on fixed relationships between the variables. A more sophisticated model would take into account the fact that the variables themselves are not independent of one another, so that for example average wages are a function of age and WLE, and population size a function of LE and birth rates. A more sophisticated model would also take into account the whole economy in which case it may be possible to show situations in which non-wage-GDP could compensate for declines in wage-GDP but this needs to be verified in further work. Nevertheless, the simple model is a useful tool for summarising how demography and the economy are linked, and how movements in their values can influence key economic indicators.

Table 8.4 is designed to show the effect on three key quantities, GDP, GDP per capita and the tax rate, of a change in model parameter by 1 year or one percentage point whilst holding all other quantities constant. It is seen that an increase in LE drags GDP down and raises taxes; whereas improvements in health and working life have the opposite effect. Increase in wages and the participation rate increase GDP and reduce taxes but increases in pensions and pensions only increase taxes. Finally, an increase in population lifts GDP only as might be expected.

Implications for UK Health and Social Policy

Based on these findings, the need to improve HLE and WLE therefore appears inescapable and so a balanced long-term approach is needed. The above scenarios reflected a time horizon of 2025 and are compared with a 2007 baseline. The

Table 8.4 Impacts of a 1 year or 1% increase in the model variables

	Effect of a +1 yr or +1% change		
Quantity	GDP	GDP/capita	Change in tax rate (%)
e_l	−1.6%	−1.6%	1.7
e_w	0	0	0.2
e_r	0	0	−0.2
e_{hw}	2.7	2.7	−2.6
e_h	0	0	−0.7
a (+1 %)	1	1	−0.3
w (+1%)	1	1	−0.3
p (+1%)	0	0	0.2
b_r (+1%)	0	0	0.04
b_w (+1%)	0	0	0.06
P20+ (+1%)	1	0	0

probability of any scenario occurring is contingent on a range of factors. For example, an increase in WLE is more difficult to achieve without accompanying improvements in HLE and labour demand; wage productivity is more difficult to sustain unless productivity of older workers in their 50s increases to levels of those in their 40s. We now work through some arguments that are barriers to improvements or opportunities that could make a difference if removed.

Increasing Healthy Life Expectancy – Barriers and Opportunities

It is reasonable to assume that LE will continue to increase at historical rates over the immediate future and to all intents and purposes it should be taken as a 'given'. The issue of prolonging lives of people who are seriously ill or disabled is an important component of life extension, as are the care consequences of the increasing numbers of centenarians expected in the next decades. One reason for concern is that male LE at age 50 is accelerating and is a key reason for supposing that current population projections will undershoot the true number of older people in years to come. This point is discussed further in a later section.

The evidence base for interventions that close the gap between LE and HLE is incomplete and fragmentary. We have not reached a stage in the state of the art where we can say that if we do x this will achieve y with an adequate degree of certainty and that progress is necessarily incremental and long-term. Claims for potential gains in health from initiatives often involve double counting of costs and sometimes exaggerated benefits for publicity effect. For example, estimates that heart disease costs the healthcare system £3.5 billion (British Heart Foundation, 2008) and stroke £2.3 billion (Department of Health, 1996) are almost certainly inaccurate due to double counting due to co-morbidity and other effects.

One reason why LE has improved so much in the last 20 years is the success of medical interventions particularly in the area of managing heart disease. This has apparently had the effect of increasing the gap between LE and HLE (i.e. years spent with disease). The complementary strategy faced with increasing LE is to increase HLE. Four main options arise: (i) improve HLE by spending more on healthcare; (ii) remove hazards in society and the work place that are known causes of ill health; (iii) promote social norms that encourage healthier lifestyles such as cohabitation and work; (iv) action on education and jobs (since these increase WLE as well as HLE). In the following sections we pick a few examples of each but these are by no means exhaustive.

Countries that spend more on healthcare generally have a higher HLE. However, studies that show gains in HLE flatten off as spend increases are based on cross-sectional data and do not take account of advances in medical technology. Nevertheless, it is interesting that a country like Japan can spend less than half the amount per capita as the USA and yet achieve an HLE of 75 years at birth as compared with 69 in the US. The UK which spends a similar amount on healthcare to Japan has an HLE of 71 based on data from the WHO. Differences in healthcare delivery, affordability, organisation and cost control are some of the underlying issues explaining the differences (e.g. countries with high proportion of private healthcare do less well), but also differences in lifestyles and degree of inequality.

Prevention is a general term used and refers to disease prevention although there does not appear to be a satisfactory way of measuring impact on HLE in a general way apart from using Quality Adjusted Life Years (QALYs). The onset of chronic disease may be regarded as inevitable in an ageing population since many other causes of death at earlier ages have fallen (e.g. accidents, infectious disease). Research shows that signs of chronic disease begin at an early age but take time to build up into a diagnosis. Once diagnosed a chronic diseases cannot be cured but can often be managed through medication and life style changes for many years.

More than 60% of all avoidable deaths are caused by cancer and cardiovascular diseases. The top ten causes of avoidable deaths are heart disease, lung cancer, suicide and self-inflicted injuries, colorectal cancer, cerebrovascular diseases, road traffic injuries, chronic obstructive pulmonary diseases (COPD), breast cancer, diabetes and alcohol-related diseases. Life expectancy from age at diagnosis of a chronic disease such as heart disease is greater than with cancer although there have been significant improvements here too. In thinking about the benefit of health interventions it is useful to distinguish between those that promote life extensions in a diseased state and those which prevent the onset of chronic disease.

Interventions that prolong life in a disease state then need to be subdivided into those that allow people to continue work (e.g. those with hypertension, diabetes) and those that might not (e.g. a stroke). The onset of chronic disease varies by individual and may be related to lifestyle or to genetic factors and so the ability to delay disease will be an issue relating to both, one of which is more amenable to change than the other. If the average age of onset of all chronic disease could be delayed by 1 year, then reasonably this might translate into a 1 year improvement in HLE and so on.

Research shows that people diagnosed with chronic disease early in life have a reasonable LE albeit in a diseased state, whereas people diagnosed with the same disease in the late stage of life have a shorter LE and consume fewer health resources over the life course. For example, a person diagnosed with COPD at age 70 has a 20% chance of dying within 3.3 years; if diagnosed at age 55, it is 9 years. The general hypothesis is that by delaying the onset of chronic conditions results in both a higher HLE and LE, but also a shorter gap in years between them.

It can be argued therefore that policies and actions that delay the onset of disease are likely to prove less costly in the long run than actions that deal with the consequences. Similar lessons were learnt in the nineteenth century in combating infectious disease through the introduction of improved sanitation. In order to measure progress in HLE at a more detailed level it would probably pay to set up a bundle of indicators to monitor age specific new cases of chronic diseases such as hypertension, diabetes etc. but definitions would need to be rigorous for comparability purposes. Reduction in the incidence of these diseases at younger ages would be one way of measuring progress towards improvement in HLE.

Fries (1980) called this process the 'compression of morbidity' and claimed that 'whether the period of morbidity is shortened depends very much on the average age of onset of the first marker (e.g. diagnosis of hypertension or first heart attack)'. The earlier the onset, the greater the likelihood of a second or third disease such as hypertension or diabetes occurring which has the effect further increasing healthcare costs through more doctor visits, prescriptions etc. (e.g. see Alder, Mayhew, Moody, Morris, & Shah, 2005). People with early stage diagnoses of one chronic disease at young ages are more likely to acquire further disease before deaths and so the burden of disease accumulates in this way and is spread out over more years.

Of all the risk factors, smoking remains the most important underlying causal factor in cases such as lung cancer, heart disease, COPD and is hence a major cause of avoidable deaths. Despite a long-term fall in adult smoking rates to around 25% today, smoking is estimated to account for around 110,000 deaths a year or around 18% of all deaths. Death from smoking related illnesses is more expensive than say death from serious stroke although there is a paucity of information on life time medical costs for different medical conditions. US research from a few years ago for example showed that life time medical costs of heavy smokers and drinkers were four times higher than for people with moderate habits (Schroeder, Showstack, & Roberts, 1979).

A key question is by how much HLE (and in turn WLE) would improve if all smoking were to stop. Unpublished research by Karlsson et al. (2007) found that no-smokers enjoyed six to seven more years of HLE than smokers, and so a complete cessation of smoking would be expected to increase HLE by 1.5 years over a period of time based on a 25% adult smoking rate. Van Baal, Hoogenveen, de Wit, and Boshuizen (2006) found that HLE in what they termed a 'healthy living cohort' was 54.8 years for men and 55.4 years for women at age 20. For male smokers, HLE was 7.8 years less and for females 6 years less. Crude calculations based on the relationship between HLE and healthcare spending show that it would require a 50% increase in health spending or about £50 billion year to achieve the same effect.

Obesity, like smoking, is another major risk factor that has an adverse impact on health, but unlike smoking, obesity is on the increase. Obesity is a condition used to describe high levels of body fat and is associated with increased risk of morbidity and mortality. The Health Survey for England shows for example that the proportion of adults classed as obese has increased in the UK from 15% in 1993 to 25% in 2006. The same survey shows that the proportion classed as morbidly obese has increased from 0.8% in 1993 to 2.1% in 2006. Obesity is associated with poor diet, reduced physical exercise and social factors as well as an increased risk of various life threatening chronic diseases. Studies have found for example that obese individuals are at increased risk of cancer, cardio-vascular diseases and diabetes and had the effect of decreasing life expectancy. Similarly the relationship between body mass index (BMI) and mortality show that risk of death increases when BMI is less than 20 kg/m^2 is optimal between 20 and 25 kg/m^2 and is increasing for BMI categories above this (Seidell, Verschuren, van Leer, & Kromhout, 1996).

A 34 year-old obese man was found to live on average 4 years less than men with healthy body fat levels and a woman 2 years less (Richardson, Mayhew, & Rickayzen, 2009). As obesity reduces the age of onset of chronic diseases, it means that HLE is reduced also but it is not known by how much. Research on the impact of obesity does not give figures for WLE but shows that that obesity exerts a large, statistically significant and negative effect on employment for both males and females after controlling for health (Morris, 2007). It appears that the negative effect is greater for the severely obese than the obese, and greater for females than males.

The other major health challenge linked to ageing is mental health problems which are also to an extent co-related with other chronic diseases especially in older age. The recent Foresight report on mental health and wellbeing in the twenty-first century is an example of another recent Government report which has expressed health concerns about the ageing population (Foresight, 2008). It notes for example that 'dementia will have a substantial and increasing impact on individuals and families with the number of people affected doubling to 1.4m in the next 30 years'. However, it also points out that mental health problems are also a factor at all ages, affecting specific sub-groups such as drug users, adolescents, the unemployed, and looked-after children.

In a Department of Health report in 2009, the cost of dementia is put at £17 billion a year and that if the onset of dementia 'could be delayed by 5 years it would reduce deaths by 30 k a year'. This is an example of how delaying the onset of a long-term condition can save lives and reduce costs. A problem is that research on how mental health problems affect LE, HLE and WLE is lacking except in obvious cases such as suicide and therefore needs further work before its full impact can be assessed (there is no such assessment in Foresight). Almost certainly the issues need to be broken down into different conditions such as dementia but also into different sub-groups to understand and measure the long-term effects (e.g. by employment status, housing tenure, household characteristics, lifestyle).

The ONS 2007 survey of adult mental health reports that the prevalence of mental health conditions requiring treatment has increased since 1993 from 14.1 to 16.4% of the adult population (National Centre for Social Research, 2009). Mental health problems overlap in part with harmful drinking habits and illicit drug taking. According to the same survey 24.2% of adults exceed the limit for non hazardous drinking and 3.8% drink harmful quantities with rates the highest in the age range 16–34. Although illicit drug use in the last 12 months is reported by 9.2% of adults this increases to 24.3% in the 16–24 age groups and 19.6% between ages 25 and 34. Evidence that mental health is an increasing problem is also provided by the increased uptake in Incapacity Benefit by people citing mental health conditions (see next section).

Increasing Working Life Expectancy – Barriers and Opportunities

Our simplified model showed that GDP per capita and GDP itself could be increased if WLE or HLE are increased. The model also showed that an increase in HLE is an important adjunct, because healthy people are more likely to be in work than unhealthy people so that strategies that promote both are more likely to be successful. There is research for example that shows that people in work enjoy better health than people out of work although clearly caveats must be applied since causation is bi-directional. However, it appears that the effect of ill health on the decision to retire is more important than the effect of retirement on ill health.

Turning to WLE, a key bottleneck within the UK labour market is the high economic inactivity rate after the age of 48 with increasing levels of disability long before state pension age is reached. According to the Labour Force Survey, of the 36.3 million people aged between 20 and 64 years 28.5 million are economically active. Of the 7.8 million economically inactive population 3.1 million are classified as Definition under Disability Discrimination Act (DDA) disabled, leaving 4.7 million who are not. Of the 3.8 million economically inactive aged between 48 and 64, 1.9 million are DDA disabled, leaving 1.9 million who are not.

Inactivity rates accelerate as state pension age is approached and it is probable that the two are associated in some way. Some of the reasons for high inactivity rates for people aged 50–59 were analysed using English Longitudinal Study of Ageing (ELSA) data. It showed that 26% of males and 28% of females had failed 1+ ADLs by their 50s and that 7.8% of males and 14% of females are carers. It was found that a male is 1.28 times more likely to work if he is educated and 2.87 times more likely if he is a home owner. In the case of females the equivalent odds are 1.72 times and 2.01 times.

Being long-term sick or disabled has a greater effect than individual caring responsibilities on work status. For example, the analysis shows that a man is 7.14 times *less* likely to work if he has failed 1+ ADLs and a woman 4.35 times. By contrast a man is 1.46 times *less* likely to work if he is a carer and a woman 1.23 times less likely. Such direct evidence suggests a filtering process in which healthy educated home owners are more likely to be economically active in their 50s even if they have caring responsibilities and poor health.

The probability of having elderly frail relatives tends to be higher in a person's 50s and so increased caring responsibilities could become a bigger barrier to work over time but health improvements could mitigate this.[12] Other evidence elicited from this analysis found for example that males were 1.4 times less likely to work if they were smokers and 2.12 times *more* likely if they were cohabiting. It was noteworthy that the same two variables had a neutral impact on females.

Health deterioration accelerates in this age range and there is a very close correlation between the LFS economically inactive disabled rate, and the percentage of people on long-term sick and disability benefits. In terms of income it is noteworthy that average weekly earnings peak when a person is in their 40s; also the number of beneficiaries of tax credits which boost income for people in work falls notably after age 50 presumably as a result of dependent children leaving home.[13]

Benefit replacement rates for people on the minimum income and average earnings and shows that for a person or couple claiming income support disability premium replacement rates are very high i.e. either income may need to be higher or benefits lower. Thus we have four factors that are affecting economic participation from an individual perspective: lower wage incentives, more caring responsibilities, increasing rates of disability, and impending state pension age.

Strong confirmation that mental health problems are replacing other conditions as a reason for economic inactivity is available from Incapacity Benefit data. The claimant load as a percentage of the working age population has increased from around 3% in the 1960s to over 7% today. However, a recent phenomenon is claimants citing mental and behavioural disorders which have increased both as a proportion of all new claimants and of the overall caseload. Those with mental and behavioural disorders as a primary indicator accounted for over 40% of the total caseload in 2006 compared to 26% in 1996. This trend represents a growing challenge as this group typically have poorer work records and prospects (Black, 2008).

Benefit data also show that the probability of leaving benefits is lower for those who have been in receipt for more than 12 months which tends to apply to older workers than those with shorter durations. Factors on the demand side of the labour equation include the difficulties of finding jobs for people 50+ that have been made redundant as a result of previous economic downturns, company closures etc., and skill gaps between jobseekers and prospective employers. The causes of economic inactivity are therefore many but the net effect of both push and pull factors has been to constrain and dampen economic activity rates in this critical age range and so prevent a crucial extension to effective WLE.

To see how slow change can be in this area we need to look at labour participation trends. Average labour participation rates over the age range reached a peak in 1990 at around 63% before falling and remaining broadly static at 62% until 2002. Since then they have started to rise again and were at 64% in 2008. This masks significant differences between males and females with rate of economic activity

---

[12] See also Carers, Employment and Services Report Series (2007), a series of reports produced by Carers UK and University of Leeds.

[13] HMRC Child and Working Tax Credits statistics December 2008, table 3.1.

among males falling from 75% at its peak in 1990 and levelling out at 71% today. The rate for females increased rapidly up to 1990 from 47% in 1984 to 52% thanks partly to the growth in part time work. Since then it has increased more slowly to around 57% in 2008.

To put these findings into a more strategic context, assuming a steady state with constant numbers of people entering the job market in their 20s, each 1% rise in participation rate would equate to around a 6 month increase in effective WLE. For males at age 20 current effective WLE is estimated to be 39 years and so an increase in participation rate of 2% over a period of time would be equivalent to an increase of 1 year. With theoretical working life expectancy of around 40 years based on state pension age for women and customary occupational retirement age for men, there is arguably room for increases in participation rates without having to increase pension age although the gap is narrow. However, built-in inertia through forced spells of inactivity and adverse employment prospects for older workers makes this theoretical limit very difficult to achieve.

LFS data on working beyond state pension age shows better news. Here participation rates increased from 8 to 11% for males and 7–12% for females between 1984 and 2008. From previous discussion, research shows that people with the longest effective WLE are educated, specialists, professionals, such as academics whose earnings tend to peak later in life and who are in better health. This suggests that investment in education and training pays off in terms of extending WLE in later career and is advantageous in finding a job after spells of absence form the labour force (e.g. to bring up children or look after elderly relatives).

Strategies Aimed at Reducing Inequalities

As well as tackling individual areas of public health concern such as smoking, obesity and excessive drinking, there is substantial research linking ill health to social inequalities and deprivation. Inequalities are defined on several different levels for an individual, neighbourhood or society and measures of inequality include income, wealth, housing, education, access to services *et cetera*. These are usually known as 'underlying causes of ill health' rather than say smoking which is a 'direct cause' and often found in more deprived areas. Outcome measures for geographical areas are usually expressed in units of excess mortality (e.g. standardised mortality ratios), or health (HLE) and there is a wealth of data that show huge variation across the country although there is as yet no targets for HLE (e.g. see *Health Statistics Quarterly*, Vol. 40, 2008).

Lifting the worst performing areas to the levels of better performing areas and thus to the level of the best, is usually how inequality targets are framed. The Government target for England is to reduce the gap in LE at birth between the fifth of local authorities with the worst health and deprivation indicators (known as 'the Spearhead Group') and the population as a whole (England), by at least 10% by 2010. This is a tall order as at the local level the differences in life expectancy can be substantial. Research carried out in Birmingham in 2008 found male life expectancy at birth in Birmingham is 76.3 years (1.25 years less for England), but that the population sub-group with the lowest life expectancy were for males in social housing and council tax band A (the lowest value band for tax purposes) (MHA Ltd, 2008). For this group the life expectancy at birth is 69.5 years, nearly 7 years less than the mean male life expectancy at birth.

In 2004–2006, the relative gap in life expectancy at birth between England and the Spearhead Group was wider than at the baseline for the target (1995–1997) for both males and females. For males the relative gap was 2% wider than at the baseline (the same as in 2003–2005), for females 11% wider (compared to 8% wider in 2003–2005) (Department of Health, 2006). To achieve the target the gap needs to be 2.32% in 2009–2011 but an examination of trends in life expectancy at national level from 1950 onwards confirms the difficulty reducing variation at the national level.

The causal mechanisms connecting inequalities to poor health are more indirect and diffuse than they are for chronic diseases but statistical associations between inequalities and poor health outcomes are convincing. Comparative European studies show that the UK has higher income related inequality than all other countries apart from Portugal (Van Doorslaer & Koolman, 2004). The Government has introduced a wider range of measures to tackle the problem and is not simply targeting life expectancy which should be regarded more as one outcome measure based on a whole raft of social policies. Briefly, they include equal opportunities legislation designed to combat gender, age, race and religious discrimination and action in areas such as child poverty, education etc., which if successfully addressed can also be expected to improve health over time.

We have already noted that international evidence suggested that HLE is improved by improving GDP and reducing inequalities. In a recession as living standards stagnate or fall, reducing inequalities becomes more important as a health stabiliser and employment for maintaining income. So the issue becomes one of whether these policies taken together will achieve improved health and at the same time be recession proof. There is no reason to suppose that they will not, but how fast and whether the actions taken will be enough is another question.

Conclusions

The UK population will age rapidly from now on as the old age support ratio (ratio of adults of working age to the population aged 65+) goes into long-term decline. This chapter finds that the implications of these demographic changes are significant and should not be underestimated. In 2007, there were 3.8 people aged 20–54 for every person aged 65+; based on official population projections this will fall to 2.8 by 2025, but it could be 2.7 if LE continues to increase at present rates. In order to restore that balance to the value in 2008 would require 14 million extra people of working age or a net population addition of 0.8 million people per year from 2008. However, if the ratio between healthy and unhealthy people is maintained, 8 million extra people are still required.

Migration, an indicator of labour shortages, has increased in recent years due in part to EU expansion and favourable economic conditions. Whereas in the 1960s there were net outflows of population, the trend has switched to net inflows currently running at around 250,000 a year. Migration is sensitive to economic factors and net inflows may fall during the present recession but the underlying labour shortages will exert significant migration pressures for the foreseeable future as the population ages. Most people would agree that population additions on the implied scale would be disproportionate and an unacceptable strain on UK resources and social structures; moreover, it would lead to its own long-term problems as migrants themselves aged.

To support the additional numbers of older people indicated from present projections will require a number of things to occur. Firstly, improvements in HLE must occur that match or preferably exceed increases in LE. Increasing HLE relative to LE will reduce the need for healthcare, older people's services, social security benefits and hence the tax burden. It will increase the pool of people available for work and enable people those who wish to work beyond retirement age to do so albeit in a more limited capacity.

As Fries has pointed out, chronic disease has become the norm in older populations and measures that can limit the age of onset of chronic disease will concentrate morbidity into fewer years and limit the increasing phenomenon of co-morbidity (multiple chronic diseases) which results in more impairment, medical care, demands for older people services *et cetera*. Spending ever more on healthcare may be self-defeating. Investing more in preventing disease may be a better investment but improved metrics are needed to measure the long-term effectiveness of prevention policies. Clearly, removing from society harmful risk factors will have wider benefits. Smoking is a classic example. More gains in health life years would be obtained from a complete cessation of smoking than would be achieved by increasing healthcare spending by 50%. However, there is a strategic weakness in this area as prevention programmes are not as well evaluated as for example are the economic benefits of new drug treatments. With the exception of a few areas, we do not know how many extra healthy life years are gained for each £1 of expenditure on prevention.

Secondly, there needs to be an increase in WLE comparable to increases in HLE. Presently, far too many people become economically inactive before normal retirement age. It is observed, for example, that people with the longest WLE tend to be educated, cohabiting, home-owning and to be healthy at age 50. Conversely, people aged 50+ are less likely to be in work if they have caring responsibilities (usually elderly relatives but also partners) or are unhealthy, a situation that applies to approximately 30% of males in this age range and 37% of females (depending on the measure and data source used). In these circumstances, changes in pension age are arguably unlikely to succeed if people vote with their feet and leave or are pushed to leave work before pension age. For many people, income before state pension age pension is topped up by working age social security benefits (Incapacity Benefit, Carers Allowance, Disability Living Allowance, Income Support, Council Tax Benefit and Housing Benefit). As pension age is increased, this will continue and expenditure will be higher unless labour participation rates are increased. One unwelcome effect of this will be to offset anticipated public expenditure gains from increasing female state pension age to 2020 and then beyond.

It is calculated that participation rates would need to increase by at least 2% in order to increase WLE by 1 year, but we calculate that the increases will need to be higher than this. Low participation rates in the 50+ age range are one of the bottlenecks identified that prevent this happening. We have not analysed labour demand issues in this chapter in detail but the fact that average wages tend to peak in a person's mid-40 s probably may lead to negative associations with work and employers and further reduce incentives to work. Benefit replacement rates start to look attractive after 50, especially for people in low paid jobs, and may provide another inducement not to undertake paid work. Labour participation rates have been slowly recovering since peaking in 1989 and are now back to the levels then. The difference is that males rates have fallen and female rates have risen. Given that the damage caused to participation rates in the past is linked to earlier recessions, it would be deeply ironic if the hard fought gains in rates in recent years were to be undermined by the current recession and thus lead to another extended period of either stagnant or falling participation rates.

In conclusion, the demography of the UK is changing rapidly and the signs are that population in the mid-2020s will exceed official forecasts. The current UK population of around 60.6 million is projected to increase to 68.9 million by 2025, and will be higher still if migration rates continue and current trends in life expectancy are maintained. To put this into a wider context, every extra million people accounted for corresponds to a city the size of Birmingham! There are hence four key economic messages from this analysis:

- If the UK is to succeed economically in the coming decades, increases in LE need to be balanced by improvements in WLE and HLE, although there is some flexibility since, to a degree, they are interchangeable.
- Failure to do so could lead to migration pressures increasing the UK population still further. To some extent higher productivity may offset these pressures but since older workers are less productive than younger workers this cannot be guaranteed.
- While a growing population will lead to greater GDP, it may not translate into improved GDP per capita, and under some scenarios living standards could fall and taxes rise steeply.
- An 'active-ageing' scenario on the other hand would result in a more manageable population, and both increased living standards and GDP growth. This would involve balanced improvements in health and working life expectancy and supply side conditions to enable people to work longer and live healthier lives.

Overall the tone of this chapter has been pessimistic in outlook. To some extent the arguments presented fly in the face of the generally received wisdom that living longer is a mark of a successful society and therefore a 'good' thing. Old age is rightly celebrated but it will not be celebrated in coming decades unless there are accompanying changes in HLE and WLE. The problem is that 1 year of extra life is being valued by society the same, whether it is a 'healthy' or 'unhealthy' life year. However, the analysis has also shown that relatively small changes in HLE and WLE can make a big difference.

A difficulty is that that HLE and WLE move very slowly over time as a result of a combination of factors in some cases acting over decades. This suggests that governments should 'proof' social policies to ensure that ones that extend LE are balanced by policies that extend HLE and WLE. The evidence of this chapter is that current policies appear to be more successful at increasing LE than they do at increasing WLE or HLE. In demographic terms, the UK is at a turning point but the real crunch is still a few years hence. This suggests there is a window of opportunity in which to change direction to one based on the 'active ageing' scenario outlined above.

In conclusion, this chapter has shown that the accumulation of healthy life years is preferable to the accumulation of unhealthy life years, but this needs to change faster if the challenges of an ageing population are to be met. A further problem is that health is measured in different ways but the metrics used in this chapter suggest that there are gradations of health and that different metrics are needed for different purposes (not all unhealthy life years are equivalent). Finally there are signs that the received wisdom that we are living longer but also living healthier for longer are also changing. As the OECD recently noted: 'it would not be prudent for policy-makers to count on future reductions in the prevalence of severe disability among elderly people to offset completely the rising demand for long-term care that will result from population ageing' (OECD, 2005).

References

Alder, K., Mayhew, L., Moody, S., Morris, R., & Shah, R. (2005). *The chronic disease burden – An analysis of health risks and healthcare usage*. London: Cass Business School.

Araujo, M. C., Ferreira, F. H. G., Lanjouw, P., & Ozler, B. (2008). Local inequality and project choice: Theory and evidence from Ecuador. *Journal of Public Economics, 92*(5–6), 1022–1046.

Black, C. (2008). *Working for a healthier tomorrow, Dame Carol Black's review of the health of britain's working age population*. London: The Stationary Office.

Blake, D., & Mayhew, L. (2006). On the sustainability of the UK state pension system in the light of population ageing and declining fertility. *The Economic Journal, 116*(512), F286–F305.

Breakwell, C., & Bajekal, M. (2005). Review of sources and methods. *Health Statistics Quarterly, 26*, 17–22.

British Heart Foundation (BHF). (2008). *Coronary heart disease statistics 2005*. http://www.heartstats.org/temp/2008.Chaptersp13%281%29.pdf

Department of Health. (1996). *Burdens of disease – A discussion document*. London: Department of Health.

Department of Health. (2006). *Health inequalities target monitoring – Life expectancy at birth*. Department of Health, London.

Department of Health (2009, February 3). *Living well with dementia: A national dementia strategy*, http://www.dh.gov.uk/en/SocialCare/Deliveringadultsocialcare/Olderpeople/NationalDementiaStrategy/index.htm

Fang, C., & Wang, M. (2006). *Demographic transition and economic growth in China*. Dewen Institute of Population and Labour Economics Chinese Academy of Social Sciences, wangdw@cass.org.cn

Foresight Mental Capital and Wellbeing Project. (2008). *Final project report – Executive summary*. London: The Government Office for Science.

Fries, J. F. (1980). Aging, natural death, and the compression of morbidity. *New England Journal of Medicine, 303*(3), 130–135.

Interdisciplinary Centre for Comparative Research in the Social Sciences (ICCR). (2005). *Overcoming the barriers and seizing the opportunities for active ageing, on behalf of the European Union* (contract No: HPSE-CT-2001-000102). http://www.iccr-international.org/activage/en/index.html

Karlsson, M. (2007, June). *Giving up smoking adds seven years to good health*. London: Cass Business School Press Release, City University.

Kawachi, I., Kennedy, B. P., Lochner, K., & Prothrow-Stith, D. (1997). Social capital, income inequality, and mortality. *American Journal of Public Health, 87*, 1491–1498.

Krugman, P. (1996). *The spiral of inequality* (pp. 44–49), Mother Jones (November/December).

MacKellar, L., Ermoliieva, T., Horlacher, D., & Mayhew, L. (2004). *The Economic impacts of ageing*. ESRI studies on Ageing, Cheltenham: Edward Elgar.

MHA Ltd. (2008). *Birmingham: Adult mortality and life expectancy*. www.nkm.org.uk

Morris, S. (2007). The impact of obesity on employment. *Labour Economics, 14*(3), 413–433.

National Centre for Social Research. (2009). *Adult psychiatric morbidity in England, 2007: Results of a household survey*. Leeds: National Health Service Information Centre for Health and Social Care.

OECD. (2005). *Long-term care for older people, the OECD health project*. Paris: OECD Publishing.

Rasulo, D., Mayhew, L., & Rickayzen, B. (2008). *Decomposition of changes in disability-free life expectancy by cause: England, 1991–2001*. http://www.uptap.net/project23.html

Richardson, J., Mayhew, L., & Rickayzen, B. (2009). *A study into the detrimental effects of obesity on life expectancy in the UK*. www.actuaries.org.uk

Schroeder, S. A., Showstack, J. A., & Roberts, H. E. (1979). Frequency and clinical description of high cost patients in 17 acute hospitals. *New England Journal of Medicine, 300*, 1306–1311.

Seidell, J., Verschuren, W., van Leer, E., & Kromhout, D. (1996). Overweight, underweight, and mortality. A prospective study of 48,287 men and women. *Archives of Internal Medicine, 156*, 958–963.

Sullivan, D. F. (1971). A single index of mortality and morbidity. *HSMHA Health Reports, 86*, 347–354.

van Baal, P. H. M., Hoogenveen, R. T. G., de Wit, A., & Boshuizen, H. C. (2006). Estimating health-adjusted life expectancy conditional on risk factors: Results for smoking and obesity. *Population Health Metrics, 4*, 14.

van Doorslaer, E., & Koolman, X. (2004). Explaining the differences in income-related health inequalities across European countries. *Health Economics, 13*(7), 609–628.

Zhao, Z. (2006). Income inequality, unequal healthcare access, and mortality in China. *Population and Development Review, 32*(3), 461–483.

Chapter 9
Spatial Microsimulation Models: A Review and a Glimpse into the Future

Mark Birkin and Martin Clarke

Introduction

In this chapter we present a review of the development of microsimulation modelling (MSM) over the past 50 years or so and attempt to outline some of the challenges and opportunities that researchers in the field are currently exploring. Phil Rees is perhaps best known for his research in fields outside MSM but, as we will indicate, he has made significant contributions largely through collaboration and supervision of research students at Leeds, so it is fitting that in this book there is a chapter that makes due acknowledgement of his work.

The chapter is structured as follows. In the next section we trace the origins of MSM from the late 1950s onwards. We note that it was not until the early 1970s that geographers picked up on the approach and began exploring the application of the models in a spatial setting. Thereafter, we review the development of this work from that time onwards. We note that putting the 'spatial' into MSM causes considerable complexity, especially as we increase the spatial resolution at which we wish to work. In the fourth section of the chapter, we examine some issues around population reconstruction within Spatial MSM (SMSM) and highlight some of the issues surrounding household dynamics within SMSM in the section that follows. We note again the complexities encountered when attempting to employ good demographic accounting standards (a point that Phil would insist upon!). The penultimate section looks at a number of interesting new developments in MSM, in particular the use of agent-based models (ABM), and attempts at embedding 'behaviour' into these models. Finally, the chapter concludes with a summary of the main points we have made in the course of the chapter and speculates about the future.

M. Birkin (✉)
School of Geography, University of Leeds, Leeds LS2 9JT, UK
e-mail: m.h.birkin@leeds.ac.uk

J. Stillwell, M. Clarke (eds.), *Population Dynamics and Projection Methods*, Understanding Population Trends and Processes 4, DOI 10.1007/978-90-481-8930-4_9,

The Early History of Microsimulation Models

As is well known, the origins of MSM can be traced back to the work of Orcutt (1957) who, with colleagues, pioneered the application of the methodology through the 1960s and 1970s (e.g. Orcutt, Caldwell, & Wertheimer, 1976). The original motivation for adopting a micro-level approach to economic modelling arose from the fact that as traditional macro-economic models are disaggregated to reflect the heterogeneity in the real world they become increasingly unwieldy, or more accurately, the occupancy matrix becomes extremely large and sparsely populated. The attraction of MSM, where individuals (people, households, firms) are represented in lists, is that as the level of disaggregation increases (the more variables that are represented), it becomes an increasingly efficient method of representing the system (for a more detailed explanation, see Clarke, Keys, & Williams, 1981).

In an early book, Orcutt, Greenberger, Rivlin, & Korbel (1961) presented the results of their efforts towards developing a model of the US socio-economic system. Although most of their efforts were directed on household dynamics, their intention was to model the economic sector as well. Despite this innovative work, MSM techniques lay largely dormant throughout the 1960s. The 1970s, however, can be seen as a time of renewed interest in MSMs. A large project under the guidance of Orcutt at The Urban Institute in Washington resulted in the development of *DYNASIM* (DYNAmic SIMulation of Income) (Orcutt et al., 1976). *DYNASIM* was a national model with no geographical representation and was used to look at a variety of policy impacts relating to the labour market. Also worthy of note around this time is the work of Kain and Apgar (1977) whose housing model at the National Bureau of Economic Research (NBER) contained a MSM to examine housing dynamics.

Since that time, MSM models have become much used in economics, particularly by policy makers at the national level, for example in the UK at HM Treasury (Roe and Rendle, 2009) and the Department for Work and Pensions (Edwards, 2009). Most of these models have been used to examine taxation and benefits impacts of various Government policies. They are almost invariably static models and have little or no geographical dimension.

Spatial Microsimulation Models

It could be argued with some justification that Torsten Hagerstrand's work in the late 1950s and 1960s (Hagerstrand, 1957) not only parallels that of Guy Orcutt in pioneering the techniques of microsimulation, but also places early emphasis on behavioural drivers with echoes of the later agent-based approaches, and does all this within a geographical context. However, the first available evidence of an operational spatial microsimulation model is the work of Wilson and Pownall (1976). This work spawned a long tradition of SMSM research at Leeds (Clarke, Keys, & Williams, 1981; Clarke, 1983; Birkin & Clarke, 1988). At the same time, interest in the use and application of SMSM grew rapidly amongst a number of research groups around the world. The great strength of microsimulation models is that they address problems of heterogeneity, averaging and aggregation bias which underpin traditional aggregate based approaches. Since spatial analysis by its very nature introduces high levels of variety and disaggregation to a population, it is not surprising that MSM has emerged as a significant sub-field for quantitative geographers.

Early applications of SMSM to problems of health care and hospitalisation have been richly extended through recent work on problems of obesity (Procter, Clarke, Ransley, & Cade, 2008), diabetes (Smith, Clarke, Ransley, & Cade, 2006) and smoking, including the integration of MSM for individual behaviour with location models for service provision (Tomintz, Clarke, & Rigby, 2008; see also Birkin & Clarke, 1987, for early proposals to fuse MSM with spatial interaction models of service choice and location). Other notable work in the early 1990s saw Phil Rees collaborating with Julia Williams on another PhD project exploring the interaction of micro-demographics and epidemiology in the context of transmission of the HIV virus. Although not widely published (Williams, 1993) this work pre-empts by at least a decade a burgeoning interest in multi-agent based individual models of epidemics and their control (e.g. Eubank et al., 2004; Ferguson et al., 2005; Epstein, 2009) and with much more plausible levels of geographical detail to boot!

Amongst the other fruitful application domains for MSM have been simulations and forecasts of water demand (Williamson, Clarke, & McDonald, 1996), social care (Williamson, 1996), labour markets (Ballas & Clarke, 2000), retailing (Nakaya, Fotheringham, Clarke, & Ballas, 2007) and education (Kavroudakis, 2009). Perhaps most notable of all is the growing attraction of MSM in the mainstream of demographic modelling and projection (van Imhoff & Post, 1998; Gampe, Zinn, Willekens, & van den Gaag, 2007), which has gone hand-in-hand with a refreshing focus on methodological enhancements, particularly regarding the development of credible dynamic MSM with spatial disaggregation. Again it is only fair to point out that Phil Rees is heavily implicated in the earliest high quality examples of dynamic spatial MSM (Duley, Rees, & Clarke, 1988). Further discussion and recent results in this strand are presented later in the chapter.

Population Reconstruction

In an ideal world, MSM would make use of available data specified at the individual and/or household level. Unfortunately, whilst micro-data sets are often available at national level (e.g. the Sample of Anonymised Records in England and Wales) rarely are such data available with sufficient geographical specification for our needs. Methods are therefore needed to reconstruct micro-level populations with the requisite spatial representation. In the early years of SMSM, the most commonly used approach was Iterative Proportional Fitting (IPF) (Clarke, 1983; Birkin & Clarke, 1988, 1989). This approach involved the creation of conditional probability distributions from aggregate tables (e.g. Small Area Statistics from the Census) and employing Monte Carlo sampling to generate lists of households and individuals for each small area, such as a census ward, in a city.

These early efforts to generate microdata from aggregate statistics were quickly superseded by reweighting methods from survey data. In the 1990s, microdata generation forming the basis for the well-known *TranSims* system at Los Alamos was performed from reweighting of the Public Use Micro Sample (PUMS) in North America, using IPF against key census distributions (Beckmann, Baggerly, & McKay, 1996). In Leeds, simulated annealing was introduced as a method for optimising the selection of small area micro-populations from survey samples (Williamson, Birkin, & Rees, 1998). This method is not conceptually difficult to grasp. In essence, the objective is to populate a small area with N individuals or households from a survey dataset of size M. Starting with a random selection of N from M, the goodness of fit of this selection can be evaluated against known population distributions (of age, household composition, ethnicity, *et cetera*) from the census or a similar source. Now we exchange a selected small area record for a selected survey record, and repeat the process as often as is necessary in order to generate better fits to the known small area demographics. The essential uniqueness of the simulated annealing method against other hill-climbing heuristics for this problem is that downhill steps are also permitted. Simulated annealing has continued to be the most popular – and probably the most effective –method for the creation of reweighted spatial micro-data, although in addition to IPF, other means to the same end include regression analysis (Tanton, McNamara, Harding, & Morrison, 2009), and genetic algorithms (Birkin, Turner, & Wu, 2006).

Nevertheless, problems and issues remain in a number of areas, in particular relating to issues of diversity and comprehensiveness. Problems associated with diversity were noted as early as 1989 in attempts to reconstruct metropolitan income distributions from a combination of census and survey sources (Birkin & Clarke, 1989), although this analysis is not straightforward, particularly since the target distribution (in this case income) is in itself unknown – that is why we are keen to use microsimulation as an estimator! In a more recent publication, Birkin and Clarke (2009) propose a method for amelioration of the difficulty using geodemographics. In effect, they argue that spatial autocorrelation undermines the naïve statistical assumptions on which reweighting is founded. For example, if one is trying to understand the variations in car ownership across a city, then ownership will tend to be low in city centres just because these are crowded, congested places with high accessibility to services and good provision of public transport. Precisely the reverse is true in rural areas. These ideas are tested against spatially tagged microdata from a survey source. One possibility here is that spatially referenced microdata could become much more widely available from either government or commercial sources. However, it is likely that concerns will remain about confidentiality of data in the public sector while issues like ownership and copyright are also important to commercial providers for whom bias is also an important consideration. Improving the diversity of spatial microsimulation models remains an important research requirement.

The robustness of microsimulations across a wide range of demographic attributes is also a subject of some concern. Consider a typical problem in which a source of individual data, for example the BHPS, is being reweighted to some aggregate small area data from the census. Typically, what will happen here is that a set of six to eight census tabulations of key variables like age, ethnicity or household composition will be chosen as a basis for determining the model 'fit'. The problem is that the BHPS contains literally hundreds of different variables, from political views to leisure pursuits. How can we have any confidence that these unconstrained attributes are reproduced reliably? The answer to this question seems to be that we cannot (cf Birkin & Clarke, 2009), and therefore we need to exercise care and judgement in the selection of our constraining variables which are appropriate to the problem at hand. So, if we are trying to build a microsimulation model to tackle the issue of obesity, then we need to choose variables which are highly correlated with obesity. This raises a number of difficulties: firstly, we do not necessarily know which variables are highly correlated with our output domain, and there will be so many interactions that selection is not straightforward; secondly, we might not know exactly the purpose of the simulation when it is initiated, and one can certainly envisage circumstances in which transferability between applications would be desirable; and finally, one would have a lot more confidence in a simulation which can represent a very wide range of personal and neighbourhood characteristics than something which can only be relied upon in a very narrow domain. Interesting work is currently in development which aims to constrain survey data across the full range of census attributes in the UK through the construction of a highly generic simulated annealing algorithm, which is capable of iterating across as many sets of constraints as the analyst cares to identify. In practice, only a limited number of univariate constraints is necessary to do this job (in the order of 20; because actually the number of source questions on the census form is not that great), and early tests indicate that the introduction of multivariate cross-tabulations for small areas does not add greatly to this process.

Household Dynamics

An interest in dynamics lies at the heart of many microsimulation models, and yet remains hugely challenging because of both technical difficulties and data deficiencies. For example, current applications include:

- *Pensim*: a model which has been constructed by the Departmernt of Work and pensions (DWP) in the UK in order to project pension requirements of the population in the medium-term future (http://www.polsim.com/PENSIM.html). Similar efforts are underway in the US Treasury Department (Holmer, 2009) and elsewhere;
- *MAP2030*: an academic enterprise as part of a multi-disciplinary project to envision the character and health and social care needs of the UK population in 2030 (http://www.lse.ac.uk/collections/MAP2030/); and
- *DynaCan*: a model which has been developed by the Statistics Canada as a basis for small area projections of the Canadian population which are extensively used in local planning of housing, transport, education and services.

Unfortunately, dynamic microsimulation modelling turns out to be hugely complex. In practice, three methods have been adopted: static ageing; transition-based simulations; and event-based simulations.

Static Ageing

The essence of the static ageing approach is that aggregate populations are projected forwards but the individual components are not. Thus, if one has a situation in which individual records are reweighted to small area (census) populations, then what happens is that some aggregate technique is used to project the small area populations forwards in time, and then the original survey data is reweighted in accordance with the new distributions. This approach has been adopted recently in the *SimBritain* project (Ballas et al., 2005). The method has an appealing simplicity, although it is not entirely clear that the implied assumption of invariance in the relationships within and between individuals and households is a robust one.

Transition-Based Modelling

Transition rate models aim to use the power of microsimulation to represent individual decision-making units as the basis for explicit probabilistic or rule-based transitions over discrete intervals of time. For example, a simple but rather powerful rule is that individuals aged x will be aged x+1 in the same geographical zone in 1 year's time, unless they migrate or die in the intervening period. The chance of migrating from one small area to another over a 12 month period is an example of a simple probabilistic transition.

An early prototype for a dynamic model with demographic transitions can be found in the work of Duley (1989) (see also Clarke, 1983). Small area demographic change was simulated using explicit models of four processes – fertility, mortality, household formation/dissolution and migration. Fertility and mortality rates were estimated directly using ONS Vital Statistics, and bi-regional migration rates were computed using patient registrations from the National Health Service Central Register (NHSCR). National rates of household formation and dissolution were derived from the ONS Longitudinal Survey (LS). The focus of the Duley model is on updating small area populations from one census to another, hence using known transition rates, extrapolated or modelled to small areas as appropriate. However, examples were also provided of the projection of baseline assumptions for small geographical areas over a 2 year forecasting period.

The concept of dynamic microsimulation with demographic transitions has recently been reviewed and extended to a generic national model in the *MoSeS* project (Birkin, Townend, Turner, Wu, & Xu, 2009).

Event-Based Modelling

The advantage of the transition rates approach is that it fits well with the data collected by many surveys and official sources. For example, census migration data in the UK are collected on the basis of 'where did you live a year ago', counts of births and deaths are usually supplied on an annual basis, and longitudinal surveys like the BHPS are also conducted at 12 month intervals. The transition rates approach also tends to mirror traditional multi-state demographic accounting approaches in which populations are seen to evolve across fixed intervals of time. A potential disadvantage of the approach is that because the micro-level populations are large and diverse, testing for transition between states could be hugely inefficient: for example, the probability of mortality in the next 12 months will typically be quite low, even for quite an elderly person, and the probability of giving birth will be quite low even for women in the highest at risk groups. Thus we can end up 'testing' for mortality, fertility and so on, quite a number of times before reaching a positive outcome.

An alternative approach to the problem which looks much more efficient computationally is to apply a test for when the next *event* is likely to take place, and then to build that event into an appropriate place in a queue of future transitions. In this case, rather than asking whether a woman of an appropriate age is likely to have a child within the next year, here we ask when a woman is expected to give birth. If the answer is 8th February 2020, then we move that individual to a place in the queue which is behind all of the other potential mothers whose children are due in the interim. This approach of simulating dynamic events within continuous time is adopted at the heart of *ModGen*, a software platform which underpins projections by Statistics Canada. A similar approach has been adopted by the *MicMac* project in the European Union (EU) (Gampe et al., 2007). The beauty of *ModGen* is that this software is now well-documented and has been freely available to the wider microsimulation community since mid-year 2009 (Statistics Canada, 2009). Whether this provides a stimulus to the further development of dynamic models within an event-based framework remains to be seen in view of the recency of these developments.

One of the ironies of the developments in SMSM is that as computational power has increased, our models have become more ambitious, more challenging but also potentially more problematic. For example, in the 1980s, household dynamic models would be typically specified at the UK census ward level, a zone that would contain in the order of 20,000 people. In 2009, the *MoSeS* model (Birkin et al., 2009) is specified at the output area level, each containing approximately 400 people. Employing Monte Carlo sampling methods at this level is not terribly robust, especially across events with a relatively low probability of happening (e.g. young adult mortality). Furthermore, the structure of UK households has grown ever more complicated. In the 1970s, the majority of households were nuclear families and people tended to live relatively close to where they worked. Now household composition is much more complex, more people are living in care homes and the student population has increased significantly. These all pose challenges for spatial population analysis as a whole and SMSM in particular.

Developments

Overview

As we have seen, microsimulation is a method with its roots in the discipline of economics. The essence of the approach is a shift in emphasis from aggregate distributions to individual 'decision-making units', thus the recognition, for example, that an increase in the gross domestic product of a country is not necessarily to the advantage of all or even the majority of its inhabitants. While the benefit of such an approach is an improved understanding of the distributional effects of behavioural or policy regimes, this obviously comes at the expense of a need for structural and behavioural models of a much wider variety of types of entities. For this reason, microsimulation models have been most conspicuously successful in contexts in which well-defined rules can be applied in different circumstances. Thus in an economic context, applications to problems involving taxation, benefits or pensions have all proven popular, in which complex but well-defined rule sets can be applied to determine the entitlements of a heterogeneous population of individuals.

In the context of demographic microsimulation, particularly with a spatial component, we can take this argument a stage further in showing that microsimulation provides a means for the efficient representation of population distributions and their associated behaviours. In other words, to improve the performance of macro-demographic models, it makes more sense after a point to completely respecify these models in micro-demographic terms than to try and disaggregate the parameters to realistic levels of detail (see Van Imhoff & Post, 1998, for a practical example in the context of fertility rate estimation.)

In addition to demographics and economics, MSM has also found applications relating to transportation and social anthropology, although transport studies in particular has carved out its own rather distinctive set of behavioural modelling approaches based around the notion of discrete choice. The development of kinship networks within an evolving population is potentially a powerful means for understanding social networks and caring relationships within an ageing society (Murphy, 2004).

While this is not necessarily the place for a detailed critique of the deficiencies of the discipline of economics in its treatment of geography, it is appropriate to observe that spatial heterogeneity adds another dimension to the variety of individual level populations, as well as adding greatly to the complexity of array-based representations of macro-populations. Thus, it is not surprising perhaps to find that geographers and regional scientists have been amongst the most enthusiastic advocates of MSM. Further distinctive applications of SMSM can be found in health, education and water demand forecasting. For example, population reconstruction techniques, as mentioned previously, have been used in separate studies to estimate the prevalence of obesity, diabetes and smoking across small areas in Leeds and Bradford (Procter et al., 2008; Smith et al., 2006; Tomintz et al., 2008). In each of these cases, the distributions are expected to vary significantly by geography, to a pattern and degree which is not captured by available data sources, and this in turn affects the demand and supply of services; for example, how many clinics are required to support people who wish to stop smoking, and where should they be located? In the case of education, then prior attainment and aspirations may vary spatially as a function of both access and socio-demographics, which may in turn affect both national and local policies on provision (Kavroudakis, 2009). Regarding the demand for water, understanding variations in consumption under alternative regimes such as unrestricted access, metering and rationing can only be achieved in relation to a detailed understanding of the composition and behaviour of individual household units (Jin, 2009).

Perhaps the greatest surprise in applied MSM is the almost universal acceptance that individuals or households are the entities of interest. From its original conception, it is clear that decision making units might equally well be conceived as, say, firms, vehicles or even buildings. As we will see below, the literature on agent-based modelling (ABM) seems much more flexible in this regard. Perhaps this is in part a function of data availability (e.g. MSM being much less abstract and more data-focused than ABM), but it would be nice to think that the future holds a more catholic vision of decision-making units within MSM (possibly through integration with ABM, a theme which we pick up later). Another divergence is in the treatment of time steps, where MSM have tended to focus on long-term strategic planning, when there seems no reason why more immediate problems, such as the incidence of disease, impact of a road closure or opening a new supermarket might not be equally well addressed.

Behavioural Models

We have seen earlier that a key weapon in the armoury of microsimulation is the notion of rules, which typically have a structure 'if x, then y'. Two important features of these rules which should be noted are that they are both rich and not necessarily deterministic. The ageing process which we referred to above is an example of a deterministic relation; thus if aged x at time t, then aged x+1 at time t+1. There are many similar examples in economic analysis, e.g. if income over £25,000 then marginal rate of taxation is 40%, and so on. However, the possibility for non-deterministic rules is even greater. For example, in a demographic model: if female and aged 18, then the probability of giving birth is q. In practice, rules are processed within a microsimulation model through a Monte Carlo sampling mechanism. In this example, we pull a random number between 0 and 1, and if this value is less than q, our (simulated) female aged 18 gives birth to a (simulated) child. The concept of richness can be illustrated if we consider a range of other factors which may affect this fertility process. For example, in addition to age and gender, then marital status, parity (including both number and spacing of previous births), ethnicity, social class, education and occupation might all be considered as potential drivers of birth rates. Such a comprehensive range of factors is well beyond the scope of traditional array-based 'macrosimulation' methods but can be handled relatively comfortably within a microsimulation framework.

An important notion that is encountered much more rarely within microsimulation models are interactions, both between individuals and between an individual and the environment. For example, if instead of the rate of taxation one is considering the price of commodities, then it may be argued that such prices are established not by rules but as the outcome of a process of interaction between buyers and sellers. The notion of interaction between individual entities lies at the heart of the new and exciting field of agent-based modelling (ABM). With its origins in computation – the notion of software agents – ABM places a very strong emphasis on the internal states and attributes of individual agents, and the evolution of these states through a process of interaction between sets of agents and their environment. A characteristic feature of this evolutionary process is that it is both top-down and bottom-up. Thus agent-based systems often exhibit the property of emergence, so that complex patterns of organisation at the environmental level can emerge naturally from the myopic actions of individual agents. Many forms of social organisation, such as markets, firms, tribes or nations, can be studied through the prism of agents (e.g. Epstein & Axtell, 1996).

Since interaction is such a crucial feature of spatial environments, ABM systems immediately suggest themselves as a powerful tool for demographics and geographical simulation. For example, recent high profile work on the transmission of both swine flu (Epstein, 2009) and Asian bird flu (HN51) (Ferguson et al., 2005) emphasises the potential of large-scale, agent-based simulations in understanding the spatial transmission of a virus, as well as the impacts of alternative policies and interventions. Nevertheless, it may also be noted that more than a decade earlier, excellent work on microsimulation of HIV/AIDS was already taking place within a MSM framework (Williams, 1993). Another good example along the same theme is that recent work on the formation of economic markets from the bottom-up (e.g. Arthur, 1999) is complemented by the even earlier work of Hägerstrand (1967) on innovation-diffusion which may be less rich in its predictive qualities but represents an approach to information-sharing based on micro-level interactions which parallels many of the interests of the ABM community.

In practice, ABM and MSM have tended to develop within distinct communities, arguably in competition with one another (although perhaps to a greater extent simply in isolation and ignorance). In general, microsimulation models are characterised by a focus on large-scale simulations, with a policy focus which is empirically grounded over long-term strategic horizons. Agent-based models are much more likely to seek theoretical insights from abstract, scaled-down representations within timescales ranging from the immediate to the evolutionary scale. However, our comments above suggest this is perhaps more a question of style and emphasis than technique. For example, large-scale epidemic simulations might just as easily come from a background in MSM as from ABM. Indeed, it might be argued that those simulations which are beginning to appear would be very much stronger from the inclusion of more realistic agent profiles and spatial behaviours which microsimulation could typically provide. In our own work, we have therefore begun to explore hybrid techniques which combine elements of the MSM and ABM approach. For example, in a demographic modelling context, we argue that such hybrid techniques can start to address difficult and fundamental challenges including the clustering of (student) minority sub-populations in urban environments (Wu, Birkin, & Rees, 2008) and widespread spatial variations in life expectancy (Wu, Birkin, & Rees, 2010). In short, we hope and expect to see many more applications in the future which transcend the divide between MSM and ABM. Perhaps in the future the common umbrella of individual based modelling (IBM) could emerge to combine the best features of each of these schools, as suggested by Wu et al. (2008).

Validation

We have argued above that much of the power and distinctiveness of microsimulation is grounded in a combination of policy-relevance and forecasting ability. In this context, it is both surprising and perhaps to some degree controversial to assert that attempts to validate the outcomes of MSM are relatively weak, and much more strongly focused towards technical checks on the robustness and consistency of procedures than assumptions and outcomes.

Regarding SMSM, then validation is a concern for both deterministic and non-deterministic models. For example, the derivation of small area populations by either synthetic generation or reweighting amounts to an attempt to estimate missing data – in effect, spatial disaggregation is performed from more aggregate distributions or constraints. In this situation, then the problem of validation poses difficulties since it is unlikely that small area estimates would be needed in the presence of complete knowledge. Two clear strategies here are to use parallel methodologies to test modelling outcomes when small area information *is* known. For example, small area demographic distributions might be used as a basis for estimating both income and car ownership in a British city. Small area income distributions are a novel output, but car ownership is known from a robust source such as the census. If simulated car ownership shows a good match to reality, then more confidence can be ascribed to local predictions for income. A second possibility would be to use sampled sources, from the Government, commercial sources or even specially commissioned investigations. The issue with both commissioned and commercial sources is likely to be heavy bias in these samples, and if this bias has a spatial manifestation then the interactions could become extremely hard to untangle. A further difficulty with Government sources is the unwillingness to provide small area identifiers for the protection of confidentiality. Major sources such as BHPS typically disaggregate only to a regional level. One possibility here could be to attach neighbourhood types, such as geodemographic codes, to such sources as suggested by Birkin and Clarke (2009) who argue that systematic variations between area types accounts for a high proportion of the errors which are typically uncovered when small area microsimulation models are validated.

When considering projections, such as in a demographic forecasting model, then the issues are more subtle still. There seems to be relatively little experience of an obvious strategy which would be to document a set of projections and then at some later date to review those projections in the light of accumulated evidence. Another approach which has been suggested but seems to be relatively untested is back projection, so that if the models are run backwards in time rather than forwards then the body of evidence for validation is much more evident! Two further issues that may be relevant in this regard are overcalibration (or possibly, in the extreme, even bogus calibration) and open source models. It is of course easy to reproduce past behaviour to acceptable levels through the introduction of sufficient parameters, although this in itself suggests that some systematic back projections could provide an interesting perspective on questions relating to model sparseness. Greater sharing and openness of simulation models could not only help to reduce duplication in the implementation of standard methodologies, but also expose the behavioural assumptions and parametrisations of MSM to beneficial scrutiny. These objectives are being pursued through a project to create a National e-Infrastructure for Social Simulation (NeISS), which is a multi-disciplinary and multi-institutional collaboration in the UK, directed from the University of Leeds. NeISS will make available neighbourhood and household or individual datasets and provide the means for not just integration, modelling, simulation and analysis but also for visualisation, publication and sharing of results (Birkin et al., 2010). It would be disappointing to see anything less than a significant acceleration of interest in the topic of validation of MSM models within the next few years.

Conclusions and Forward Look

So what can we expect the future to hold for spatial microsimulation? Much of what we suggest is perhaps implied to a greater or lesser extent by our previous observations, but for that reason alone provides a useful summary and conclusion to this review.

As far as applications are concerned, then the range of existing applications is already impressive. That said, there is still a huge bias in emphasis towards financial and economic applications. For example, at the second conference of the recently instituted International Microsimulation Association (IMA) in Ottawa, Canada (http://www.statcan.gc.ca/conferences/ima-aim2009/index-eng.htm), an eclectic group of contributors from over twenty countries contributed more than 200 papers, but the vast majority of these espoused an essentially economic theme, with strong emphases on housing, taxation and benefits, employment and pensions. It should be noted here, perhaps, that microsimulation has been and should continue be widely adopted at a global scale, with particular strongpoints across Western Europe – especially the UK, France, Sweden; in both Canada and the US, and significant activity in Australia. The next 10 years will likely see explosive growth in China (Li, Mao, Zeng, & Wang, 2008), while the relatively paucity of interest in Japan is perhaps surprising and might be corrected.

We hope and expect to see a more balanced range of applications in the future. In particular, interest in health care and epidemiological uses of MSM will be critically important. This looks increasingly like a 'killer application' for SMSM given the way that social and spatial interactions within a geographically heterogeneous population combine in rich ways to mediate the transmissions of diseases and epidemics; and given the huge investments in drug stockpiles, and the political ramifications of infection control. We expect that such concerns might also accelerate a coming together of the MSM and ABM approaches with more wide-ranging individual-based models of human behaviour (cf. Wu et al., 2008). Such developments would also provoke a greater interest amongst the MSM community in short-range planning problems alongside the more typical long-range strategic concerns.

From the preceding remark, a comment about engagement of spatial MSM in the planning process is in order. One of the great strengths of MSM has been a demonstrable value within applied planning contexts, but again it is within the financial and economic sector that this has been seen most clearly. It is certainly to be hoped that such benefits will be seen more clearly in areas like retail analysis, land use planning and demographic analysis. For example, retailers are increasingly overwhelmed by individual level customer data, which would provide a perfect jumping off point for simulations of marketing strategy, brand management and network development. This comment in itself perhaps exposes the potentially crucial argument that the increasing 'deluge' of individual level data will surely fuel a growing interest in techniques like SMSM which are capable of representing and exploiting these sources (cf Savage & Burrows, 2007, for a sociological spin on this debate). Statistics Canada has now adopted a dynamic SMSM for its core demographic projections, and so perhaps this approach might find increasing favour in other countries such as the USA and UK. A broader portfolio of policy relevant applications would also inevitably consolidate the requirement for proof of benefit (i.e. the validation question) and expose new challenges requiring refinement and theoretical development of model capabilities.

An impetus to more widespread adoption of MSM techniques, both spatial and otherwise, could be provided by greater availability of tools, data and methods. Once again, Statistics Canada are leading the way here in their promotion of the *ModGen* software as a freely available and generic engine for dynamic simulation. Unfortunately the event-based representation of transitions looks to be deeply embedded in the *ModGen* system. The extent to which this is an impediment to widespread uptake remains to be seen. Other commentators have clearly identified the need for some kind of repository of MSM models and methods as a key precursor to accelerated development of both theory and applications (Williamson, 2010). An attempt to establish a version of such a repository as a communal e-infrastructure is now underway (Birkin et al., 2010).

References

Arthur, W. (1999). Complexity and the economy. *Science, 284*, 107 –109.

Ballas, D., & Clarke, G. (2000). GIS and microsimulation for local labour market policy analysis. *Computers, Environment and Urban Systems, 24*, 305 –330.

Ballas, D., Clarke, G., Dorling, D., Eyre, H., Thomas, B., & Rossiter, D. (2005). SimBritain: A spatial microsimulation approach to population dynamics. *Population, Space and Place, 11*(1), 13 –34.

Beckmann, R., Baggerly, K., & McKay, M. (1996). Creating synthetic baseline populations. *Transportation Research, 30A*(6), 415 –435.

Birkin, M., & Clarke, G. (2009, June 8 –10). *Improving the representation of diversity in spatial microsimulation models*. Ottawa: IMA.

Birkin, M., & Clarke, M. (1987). Comprehensive models and efficient accounting frameworks for urban and regional systems. In D. Griffith & R. Haining (Eds.), *Transformations through space and time* (pp. 169 –195). The Hague: Martinus Nijhoff.

Birkin, M., & Clarke, M. (1988). Synthesis: A synthetic spatial information system for urban modelling and spatial planning. *Environment and Planning A, 20*, 1645 –1671.

Birkin, M., & Clarke, M. (1989). The generation of individual and household incomes at the small area level using synthesis. *Regional Studies, 23*, 535 –548.

Birkin, M., Procter, R., Allan, R., Bechhofer, S., Buchan, I., Finch, J., et al. (2010). The elements of a computational infrastructure for social simulation. *Philosophical Transactions of the Royal Society Series A, 368*, 3797 –3812.

Birkin, M., Townend, P., Turner, A., Wu, B., & Xu, J. (2009). MoSeS: A Grid-enabled spatial decision support system. *Social Science Computer Review, 27*, 493–508.

Birkin, M., Turner, A., & Wu, B. (2006). A synthetic demographic model of the UK population: Methods, progress and problems. *Proceedings of the second international conference on e-social science*, Manchester.

Clarke, M. (1983). *Integrating dynamical models of urban spatial structure and activities: An application to urban retail systems*. Unpublished PhD thesis, University of Leeds, Leeds.

Clarke, M., Keys, P., & Williams, H. C. W. L. (1981). Microsimulation. In N. Wrigley & R. J. Bennett (Eds.), *Quantitative geography*. London: Routledge and Kegan Paul.

Duley, C. (1989). *A model for updating census-based household and individual information for inter-censal years*. Unpublished PhD thesis, School of Geography, University of Leeds, Leeds.

Duley, C., Rees, P., & Clarke, M. (1988). *A microsimulation model for updating households in small areas between censuses* (Working Paper 515). School of Geography, University of Leeds, Leeds.

Edwards, S. (2009, April 2). *Developing and maintaining microsimulation models at DWP*. London: ESRC Microsimulation Series.

Epstein, J. (2009). Modelling to contain pandemics. *Nature, 460*, 687.

Epstein, J., & Axtell, R. (1996). *Growing artificial societies: Social science from the bottom up*. Washington, DC/London: Brookings Institution Press/MIT Press.

Eubank, S., Guclu, H., Anil Kumar, V., Marathe, M., Srinivasan, A., Toroczkai, Z., et al. (2004). Modelling disease outbreaks in realistic urban social networks. *Nature, 429*(6988), 180 –184.

Ferguson, N. M., Cummings, A. T., Cauchemez, S., Fraser, C., Riley, S., & Meeyai, A. (2005). Strategies for containing an emerging influenza pandemic in Southeast Asia. *Nature, 437*(7056), 209 –214.

Gampe, J., Zinn, S., Willekens, F., & van den Gaag, N. (2007). Population forecasting via microsimulation: The software design of the MicMac project. *Eurostat: Methodologies and working papers; Theme: Population and social conditions* (pp. 229 –233). Luxembourg: Office for Official Publications of the European Communities.

Hagerstrand, T. (1957). Migration and area. In D. Hannerberg, T. Hagerstrand, & B. Odeving (Eds.), *Migration in Sweden*, Lund Studies in Geography, Series B, *13*, 27 –158, Lund.

Hägerstrand, T. (1967). *Innovation diffusion as a spatial process,* Chicago: University of Chicago Press.

Holmer, M. (2009 8 –11 June). Projecting future US pension benefits, Ottawa: IMA.

Jin, J. (2009). *Microsimulation and water demand forecasting*. Unpublished PhD thesis, School of Geography, University of Leeds, Leeds.

Kain, J. F., & Apgar, W. C. (1977). *Simulation of the market effects of housing allowances, baseline and policy simulations for Pittsburgh and Chicago* (Vol. II). Cambridge, MA: Department of City and Regional Planning, Harvard University.

Kavroudakis, D. (2009). *Dynamic spatial microsimulation modelling of education in UK.* Unpublished PhD thesis, School of Geography, University of Sheffield, Sheffield.

Li, X., Mao, W., Zeng, D., & Wang, F. (2008). *Agent-based social simulation and modeling in social computing*. Lecture Notes in Computer Science, 5075/2010. Berlin: Springer.

Murphy, M. (2004). Tracing very long-term kinship networks using SOCSIM. *Demographic Research, 10*(7), 171 –196.

Nakaya, T., Fotheringham, A. S., Clarke, G., & Ballas, D. (2007). Retail modelling combining meso and micro approaches. *Journal of Geographical Systems, 9*, 345 –369.

Orcutt, G. (1957). A new type of socio-economic system. *Review of Economics and Statistics, 58*, 773 –797.

Orcutt, G., Greenberger, M., Rivlin, A., & Korbel, J. (1961). *Microanalysis of socioeconomic systems. A simulation study*. New York: Harper and Row.

Orcutt, G., Caldwell, S., & Wertheimer, R. (1976). *Policy exploration through micro-analytical simulation*. Washington, DC: The Urban Institute.

Procter, K., Clarke, G., Ransley, J., & Cade, J. (2008). Micro-level analysis of childhood obesity, diet, physical activity, residential socio-economic and social capital variables: Where are the obesogenic environments in Leeds? *Area, 40*(3), 323 –340.

Roe, D., & Rendle, D. (2009, September 11). *Microsimulation in the treasury: Methods and challenges*. Brighton: BSPS.

Savage, M., & Burrows, R. (2007). The coming crisis of empirical sociology. *Sociology, 41*, 885 –899.

Smith, D. M., Clarke, G. P., Ransley, J., & Cade, J. (2006). Food access & health: A microsimulation framework for analysis. *Studies in Regional Science, 35*(4), 909 –927.

Statistics Canada. (2009). *ModGen developers guide*. Online at http://www.statcan.gc.ca/microsimulation/modgen/modgen-eng.htm

Tanton, R., McNamara, J., Harding, A., & Morrison, T. (2009). Small area poverty estimates for Australia's Eastern Seaboard in 2006. In A. Zaidi, A. Harding, & P. Williamson (Eds.), *New frontiers in microsimulation modelling* (pp. 79 –96). Farnham: Ashgate.

Tomintz, M., Clarke, G., & Rigby, J. (2008). The geography of smoking in Leeds: Estimating individual smoking rates and the implications for the location for stop smoking services. *Area, 40*(3), 341 –353.

van Imhoff, E., & Post, W. (1998). Microsimulation methods for population projection. *Population: An English Selection, 10*, 97 –138.

Williams, J. (1993). *A simulation model of the diffusion of HIV and AIDS in the United Kingdom.* Unpublished PhD thesis, School of Geography, University of Leeds, Leeds.

Williamson, P. (1996). Community care policies for the elderly, 1981 and 1991: A microsimulation approach. In G. Clarke (Ed.), *Microsimulation for urban and regional policy analysis* (pp. 64 –87). London: Pion.

Williamson, P., Birkin, M., & Rees, P. (1998). The estimation of population microdata using data from small area statistics and samples of anonymised records. *Environment and Planning A, 30*, 785 –816.

Williamson, P., Clarke, G., & McDonald, A. (1996). Estimating small area demands for water with the use of microsimulation. In G. Clarke (Ed.), *Microsimulation for urban and regional policy analysis* (pp. 117 –148). London: Pion.

Williamson, P. (2010). *Microsimulation: Bridging the gaps*. Brighton: BSPS. September 11, online at http://pcwww.liv.ac.uk/~william/ESRC-BSPS%20MSM%20Seminar%20Series_files/Williamson2%20edited%20transcript.pdf

Wilson, A., & Pownall, C. (1976). A new representation of the urban system for modelling and for the study of micro-level interdependence. *Area, 8*, 256 –264.

Wu, B., Birkin, M., & Rees P. (2008). A spatial microsimulation model with student agents. *Computers Environment and Urban Systems, 32*, 440 –453.

Wu, B., Birkin, M., & Rees, P. (2010). A dynamic microsimulation model with agent elements for spatial demographic forecasting. *Social Science Computing Review*. Published online first on May 18, 2010, DOI 10.1177/0894439310370113.

Chapter 10
Tracking the Evolution of the Populations of a System of Cities

Alan Wilson and Joel Dearden

Introduction

The exploration of the evolution of systems of cities has a long history – illustrated by Berry's (1964) classic paper. In this chapter we take a simple model that is usually used for modelling the evolution of retail centres within a city and re-interpret it as a model of a system of cities. The retail model is outlined initially and its 'system' interpretation thereafter. We then explain the idea of urban 'DNA' and its evolution. This is followed by a description of the current system of interest – the evolution of Chicago from 1790 to 1870 in the context of the development of the United States in that period with particular reference to railways. Some results are presented that explore the evolution of the populations of cities in this system and some of the many possible avenues for further research are discussed in the concluding section.

The Retail Model

The simple aggregated retail model can be constructed as follows. Define $S_{ij}(t)$ as the flow of spending from residents of i to shops in j at time t. We add a similar t label to the remaining definitions to prepare the way for building a dynamic model. Let $e_i(t)$ be spending per head and $P_i(t)$ the population of i. The parameter $c_{ij}(t)$ is the travel cost from zone i to zone j at time t. $W_j(t)$ is a measure of the attractiveness of shops in j which, for these illustrative purposes, we take as the logarithm of 'size' – reflecting range of choice and lower prices through scale economies. The vector $\{W_j(t)\}$ can then be taken as a representation of urban structure at time t – the configuration of W_js. If many W_js are non-zero, then this represents a dispersed system. At the other extreme, if only one is non-zero, then that is a very centralised system. There is clearly, potentially, a measure of order in this specification of structure. The obvious order parameter (Wilson & Dearden, 2010) would be $N(W_j > 0)$ – the number of centres which are non-zero. In a fully dispersed system, then

A. Wilson (✉)
Centre for Advanced Spatial Analysis, University College London, London, UK
e-mail: a.g.wilson@ucl.ac.uk

J. Stillwell, M. Clarke (eds.), *Population Dynamics and Projection Methods*, Understanding Population Trends and Processes 4, DOI 10.1007/978-90-481-8930-4_10,

$N(W_j > 0)$ would be equal to the number of possible centres and would be large; while in a very centralised system, $N(W_j > 0)$ would be 1.

A spatial interaction model can be built by maximizing an entropy function in the usual way (Wilson, 1967, 1970). We add a time label so that the model can be used within a dynamic context below. Then:

$$S_{ij}(t) = A_i(t)\, e_i(t)\, P_i(t)\, W_j(t)^{\alpha(t)}\, e^{-\beta(t)c_{ij}(t)} \tag{10.1}$$

where:

$$A_i(t) = 1 \Big/ \sum_k W_k(t)^{\alpha(t)}\, e^{-\beta(t)c_{ik}(t)} \tag{10.2}$$

to ensure that:

$$\sum_j S_{ij}(t) = e_i(t)\, P_i(t) \tag{10.3}$$

and:

$$\sum_{ij} S_{ij}(t) \log W_j(t) = X(t) \tag{10.4}$$

where $\log W_j(t)$, as we noted earlier, is taken as the measure of consumer benefits and X(t) an estimate of the total benefits achieved. We also have:

$$\sum_{ij} S_{ij}(t) c_{ij}(t) = C(t) \tag{10.5}$$

$\alpha(t)$ and $\beta(t)$ are parameters [actually, the Lagrangian multipliers associated with Equations (10.4) and (10.5)]. Because the matrix is only constrained at the origin end, we can calculate the total flows into destinations as:

$$D_j(t) = \sum_i S_{ij}(t) = \sum_i \left[\frac{e_i(t)\, P_i(t)\, W_j(t)^{\alpha(t)}\, e^{-\beta(t)c_{ij}(t)}}{\sum_k W_k(t)^{\alpha(t)}\, e^{-\beta(t)c_{ik}(t)}} \right] \tag{10.6}$$

A suitable hypothesis for representing the dynamics is (Harris & Wilson, 1978):

$$\Delta W_j(t, t+1) = \varepsilon(t) \left[D_j(t) - K(t)\, W_j(t) \right] \tag{10.7}$$

where $K(t)$ is such that $K(t)W_j(t)$ can be taken as the (notional) cost of running the shopping centre in j. This equation then says that if the centre is profitable, it grows; if not, it declines. The parameter ε determines the speed of response to these signals. The equilibrium position is given by:

$$D_j(t) = K(t)\, W_j(t) \tag{10.8}$$

which can be written out in full as:

$$\sum_i \left[\frac{e_i(t)\, P_i(t)\, W_j(t)^{\alpha(t)}\, e^{-\beta(t)c_{ij}(t)}}{\sum_k W_k(t)^{\alpha(t)}\, e^{-\beta(t)c_{ik}(t)}} \right] = K(t)\, W_j(t) \tag{10.9}$$

and these are clearly nonlinear simultaneous equations in the $\{W_j(t)\}$.

The dynamics are given by:

$$W_j(t+1) = W_j(t) + \Delta W_j(t, t+1) \tag{10.10}$$

If the populations and the total floor space do not change, then Equations (10.7) and (10.10) simply represent moves towards equilibrium.

Re-interpreting and Extending the Model for a System of Cities

We now assume that each node in the system is a city (or a town or a village) with population, P_i and level of economic activity, W_i. e_i is the average level of economic activity generated per capita and S_{ij} represents the level of interaction between places – to be interpreted mainly as trade flows – but with an accompanying implicit assumption, that we will formulate formally below, that migration follows trade. c_{ij} is a measure of transport cost as usual and K_i is a measure of the cost per unit of maintaining a level of economic activity at i. We will have the possibility of e_i representing a spectrum from poor to rich, and similarly, K_i, from cheap to expensive, and so in a sense reflecting 'rent'.

A particularly interesting development is to examine the dynamics of $\{c_{ij}\}$ and here we do this exogenously using an underlying spider network. It is interesting and important to do this as an extension of the usual model but also because it may be critical to the evolution of the system around Chicago.

For the total population, for illustration, we can assume an annual rate of increase – say from net migration and births over deaths, of, say, λ_t, from t to t+1 for each t. We can obviously vary this assumption. We also ought to introduce some 'noise' into the system. However, the key assumption we make is that the change in population at i is determined by the change in the level of economic activity. We can combine these assumptions as follows:

$$P_i(t+1) = \mu(t)\{P_i(t)[1+\phi_{1i}] + \phi_2 \Delta W_i(t, t+1)\} \tag{10.11}$$

where ϕ_{1i} is a random variable, suitably scaled, with a mean considerably less than 1, ϕ_2 is a constant which represents the scale of population change related to change in economic activity (which can be positive or negative) and $\mu(t)$ is a normalising factor to ensure that the overall growth rate is λ_t. Hence:

$$\sum_i P_i(t+1) = (1+\lambda_t) \sum_i P_i(t) \tag{10.12}$$

so that $\mu(t)$ is determined from:

$$\mu(t) \sum_i \{P_i(t)[1+\phi_{1i}] + \phi_2 \Delta W_i(t, t+1)\} = \sum_i (1+\lambda_t) P_i(t) \tag{10.13}$$

and hence:

$$\mu(t) = \frac{\sum_i (1+\lambda_t) P_i(t)}{\sum_i \{P_i(t)[1+\phi_{1i}] + \phi_2 \Delta W_i(t, t+1)\}} \tag{10.14}$$

The model would then be run by working through Equations (10.1), (10.2), (10.6), (10.7), (10.9)[1] and (10.10); and then adjusting $\{P_i\}$ through Equation (10.11), and recycling through Equations (10.1), (10.2), (10.6), (10.7), (10.9) and (10.10) with $P_i(t+1)$.

System 'DNA' and Its Evolution

It is well known that the pattern of evolution of a dynamical system – the core of the dynamical model in our case being Equations (10.7), (10.10) and (10.11) – are strongly dependent on the initial conditions. In the case of urban and regional systems, it has been argued (Wilson, 2008) that for each step in the evolution of a dynamical system represented by these kinds of difference equations, the 'initial conditions' at time t, as determinants of the equations solutions at time t+1, can be regarded as the 'DNA' of the system. This is because since the changes in a step are likely to be relatively small, the possibilities of change – what might be called the 'cone of possible development' – will be strongly determined by what is there. This accords with a common sense view of the situation: that the existing infrastructure, economic activities and populations – will determine what is possible in the immediate future. In the model presented here, therefore, the 'DNA' can be taken as a string – the set of scalars, vectors and matrices:

$$[\{e_i\}, \{P_i\}, \{W_j\}, \{c_{ij}\}, \{K_i\}, \alpha, \beta, \lambda_t, \varphi_1, \varphi_2] \tag{10.15}$$

The i-elements:

$$[e_i, P_i, W_i, c_{ij} \text{ (for all } j), K_i, \alpha, \beta, \lambda_t, \varphi_1, \varphi_2] \tag{10.16}$$

can be taken as the 'DNA' characteristics of the zone and can be used to build typologies.

As the system evolves over time, the DNA also evolves. This represents the slow dynamics of the system. It is the $\{S_{ij}\}$ that represents the fast dynamics of the system – the 'physiology'. It is already clear that in the model, some of the DNA variables are exogenous and some are endogenous. The model will predict the evolution of the endogenous elements with the time lines of the other elements

[1]Implying that we run the W_j system to equilibrium using an inner loop every iteration of the model.

specified outside the model. The ambition in model development is always to extend the list of endogenous variables and to minimise the extent of the exogenous ones.

The model can in principle be used in three ways: first, to seek to explain the evolution of a system of interest – and this is where it is potentially of interest to historical geography. Secondly, to establish typologies based on the DNA. Thirdly, in a contemporary planning context to explore ‘genetic medicine’: how can the DNA string be modified to take the system to a desired path outside the cone of possible development? In the rest of the paper, we focus on the first of these objectives and take as our system of interest the evolution of the regional system around Chicago from 1790 through the nineteenth century.

The System of Interest

Chicago and its wider environs makes an excellent case study and there is a very good history (Cronon, 1991) on which we have relied heavily. Population census data is available for each decade from 1790 and in our model each iteration represents 1 year. There is rapid development much influenced by changes in transport technology, notably the building of railway lines. Since we are interested in both the opening up of the mid-west and the access to markets in the north-east, we consider a large region for our system of interest as shown in Fig. 10.1.

The data (or the assumptions when data is not available) for the model runs is assembled as follows. Population data (P_i) is obtained from the NHGIS census

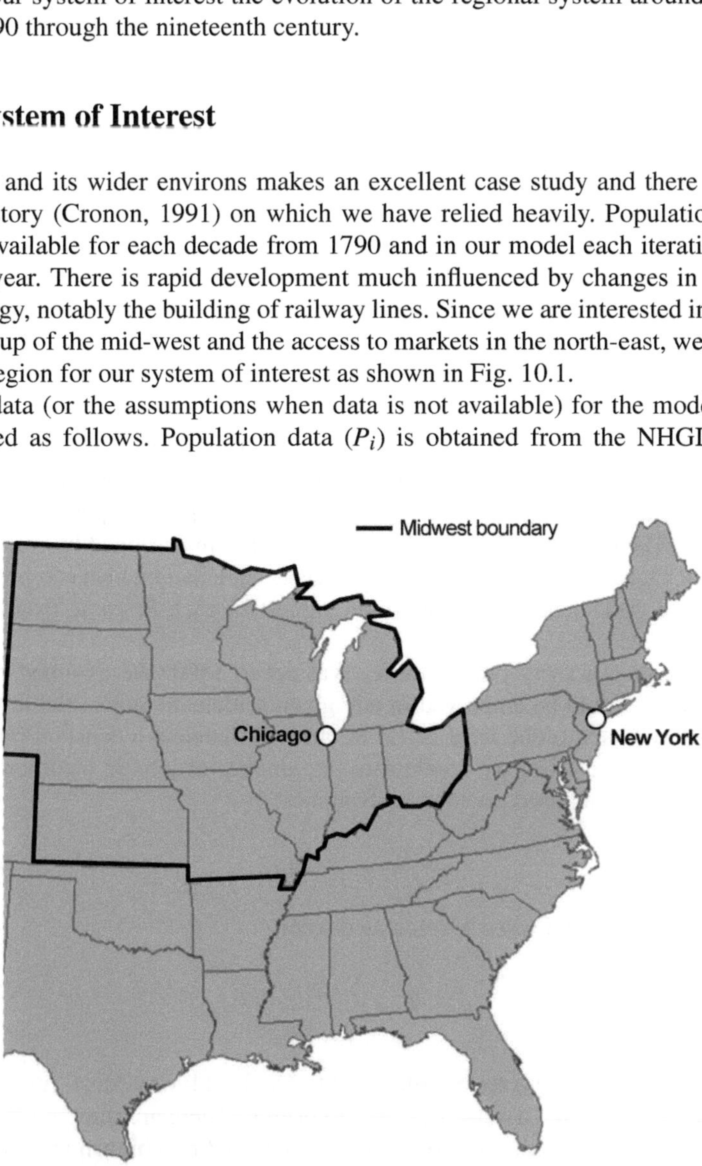

Fig. 10.1 Model area with Midwest boundary

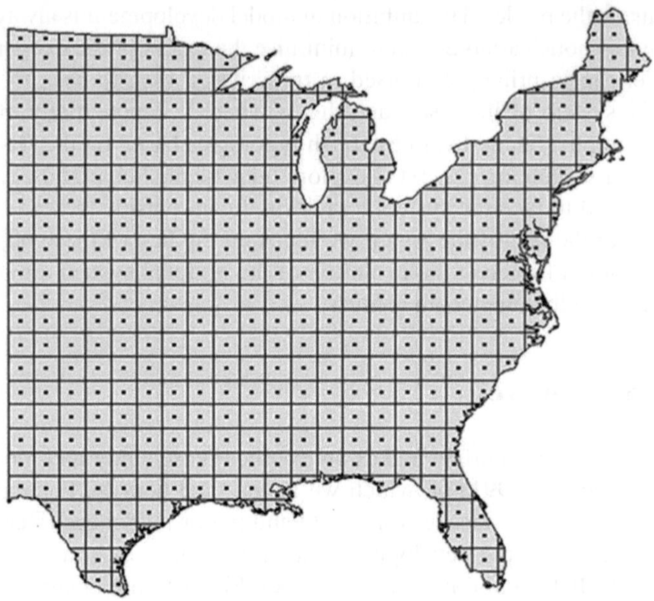

Fig. 10.2 The grid of aggregated settlements

website at www.nhgis.org. We used county data collected every decade of our study period, and since the county boundaries changed on a regular basis, we aggregated the data to a 120 km square grid to achieve period by period consistency (Fig. 10.2). The centroid of each grid square (after being cropped by water boundaries) is then taken as the position of an aggregate settlement which is representative of the whole grid square area. The initial values for each P_i come from the 1790 census.

The economic activity per capita, e_i, was set to \$500. We assumed a third of the population would be employed in any given settlement and so the initial value for each W_j was set to one third of P_i. In order to maintain a constant ratio of 3:1 in population versus jobs we recalculate the number of jobs in a settlement after calculating the normalised population dynamics:

$$W_i(t) = P_i(t)/3 \tag{10.17}$$

K_i was set to a constant value K calculated from:

$$K = \frac{\sum_i e_i P_i}{\sum_i W_i} \tag{10.18}$$

The transport costs (c_{ij}) are calculated on the basis of lowest 'cost' routes through a spider network using Dijkstra's (1959) algorithm. There are three kinds of links: roads, water and rail, with rail links further subdivided into branch and trunk lines.

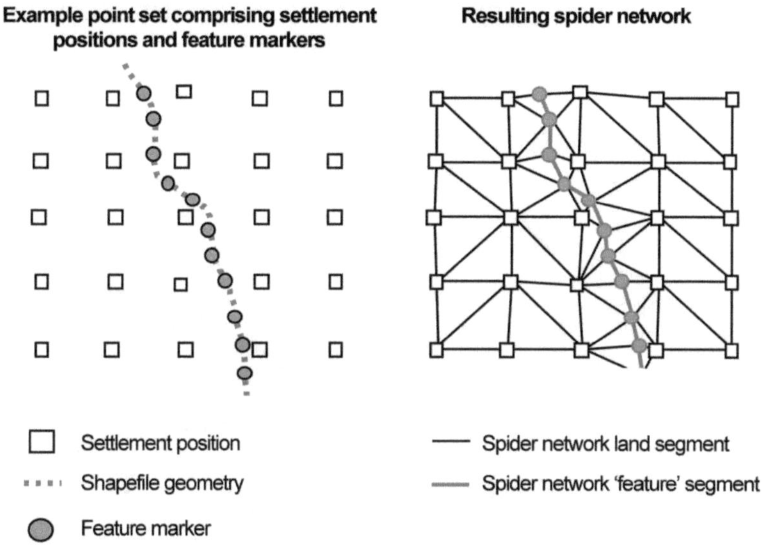

Fig. 10.3 A method for adding shapefile features into the spider network

We upgrade specific links at the appropriate iteration in a model run to represent railway construction. The links are weighted through three parameters: if road distance in km is the unit, then a water link is multiplied by *w*, say, a branch rail link by *r* and trunk lines by *t*. In the tests presented below, *w* is taken as 1/16 and *r* as 1/20 and *t* as 1/40. We construct the spider network by calculating a Delaunay (1934) triangulation of the aggregated settlement positions together with additional points that mark the path of water and railways. These additional feature marker points are derived from shapefiles of appropriate data (i.e. rivers, lakes, coastline and railways) and the general method, applicable to all kinds of data, is illustrated in Fig. 10.3.

Each feature marker was given a type indicating the kind of feature it represented, either water or rail. Railway feature markers additionally contained data on year of construction. The network links produced by the Delaunay triangulation were given their type based on the rules which took into account the type of the two end points and the position of the link in relation to the original shape file features. The spider network around Chicago is shown in Fig. 10.4 and the full spider network is shown in Fig. 10.5

The model parameters α, β, ε, φ_1, φ_2, *w*, *r* and *t* were estimated by a combination of manual and automated calibration which aimed to maximise the average r^2 value for the whole 80 year period when comparing the model output to the census data for the appropriate year – in order to obtain year by year census records we interpolated between the nearest two decades. The national increase in population for each iteration λ_t was also calculated from the interpolated year by year census records.

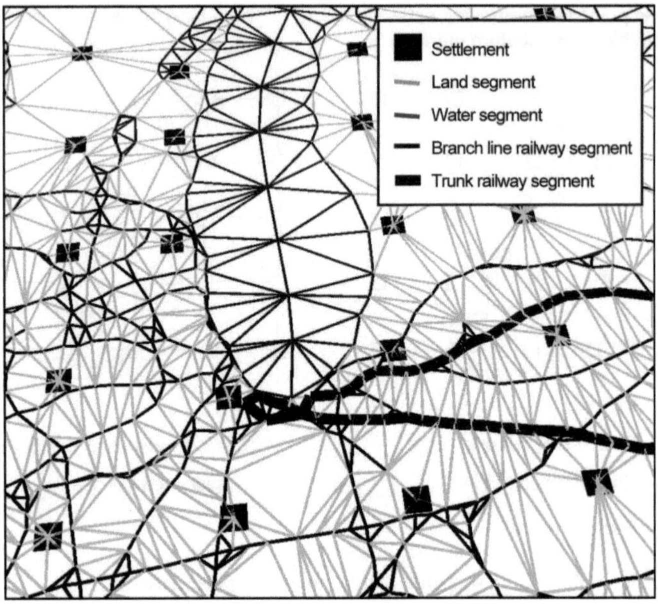

Fig. 10.4 Spider network around Chicago and Lake Michigan

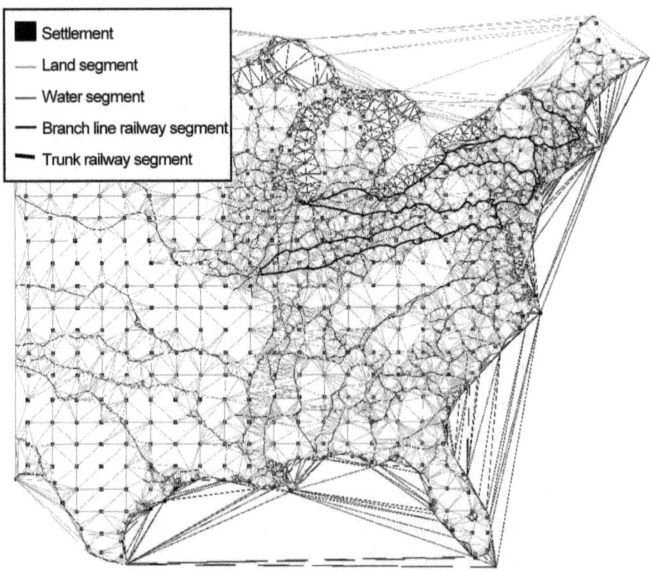

Fig. 10.5 The complete spider network for 1870 showing all rail lines

Results

The best fit model parameters ($\alpha = 0.26$; $\beta = 0.03$; $\varepsilon = 0.0006$; $\varphi_1 = 0.01$; $\varphi_2 = 3$; $w = 0.06$; $r = 0.05$; $t = 0.025$) gave an average r^2 over all 80 years of ~0.24. The Chicago grid square population was 264,546 compared to 456,959 in the census data giving an error of –192,413. Figure 10.6 shows four evenly spaced years from the best fit model run. From 1791 to 1840, growth in the model is mainly confined to the east coast, with some small settlements also appearing along the south coast. Here the model output is less concentrated in the north than the census data and this may be because we do not model the influence of sea ports such as New York – a possible future model improvement. In the early 1840s, we see growth following the construction of the railway lines westwards and this intensifies so that by 1870 a new transport corridor exists between New York and the Midwest along which large settlements have developed representing a major change to the distribution of economic activity in the country. The model also offers estimates of the trade flows

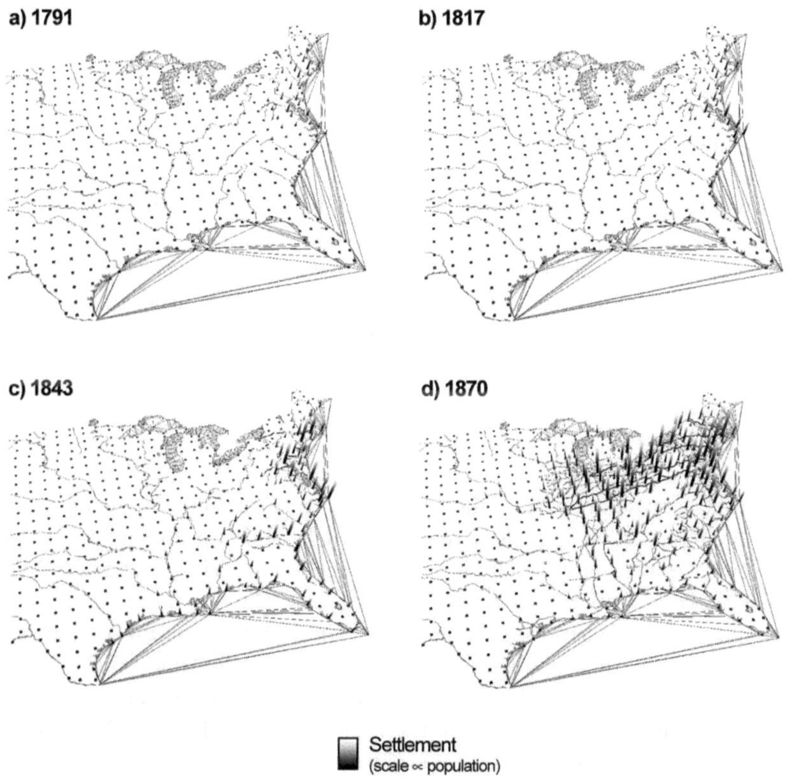

Fig. 10.6 Best fit model output

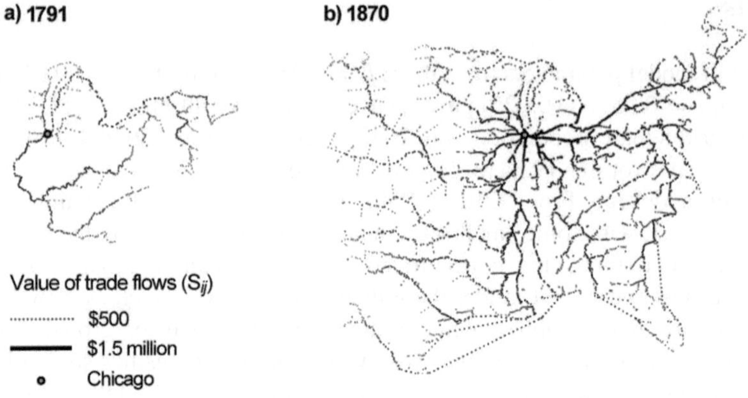

Fig. 10.7 Predictions of export flows from Chicago in (**a**) 1791 and (**b**) 1870

between zones and in Fig. 10.7 we show the flows from Chicago for 1791 and 1870. Figure 10.8 provides a comparison of the model population and census data for each decade between 1790 and 1870.

Further Developments

There are many possible improvements that can be explored with this type of model in future work and we conclude by outlining some of these. In the next two sections, we explore disaggregation and we indicate how the attractiveness factor could be expanded. Thereafter, we explore the challenge of modelling transport system dynamics and offer a concluding comment to finish.

Disaggregation

In the model as presented, W_j is taken as an aggregate measure of economic activity. In practice, there are at least four markets which make up this aggregate and we need to explore at some point treating these separately. They are: grain, livestock and meat packing, lumber, and manufactured goods. The first three are exports from the mid-west, particularly to the north-east. The last is made up of exports from the north-east some of which are imports to the mid-west. This last is complicated by the fact that much of the distribution to small towns and settlements was by the newly-invented mail order. If the volumes in each case could be translated into value, then there would be common units to produce an aggregate value. Alternatively, the model could be run for the four markets separately or re-aggregated as indicated in the next subsection through the attractiveness function.

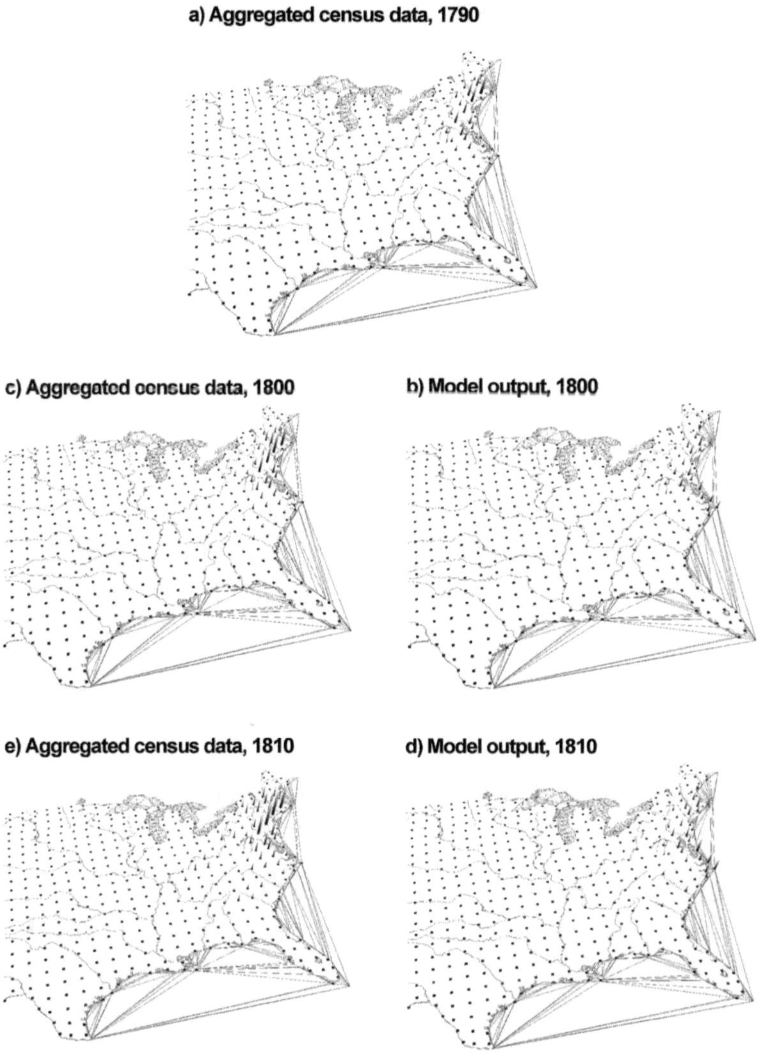

Fig. 10.8 (continued)

Expanding the Attractiveness Factor

There are at least two ways of disaggregating the attractiveness function (and indeed these two ways could be used in combination). In each case, W_j would be broken down into a series of multiplicative factors:

$$W_j^{\alpha} = W_j^{(1)^{\alpha(1)}} W_j^{(2)^{\alpha(2)}} W_j^{(3)^{\alpha(3)}} W_j^{(4)^{\alpha(4)}} \tag{10.19}$$

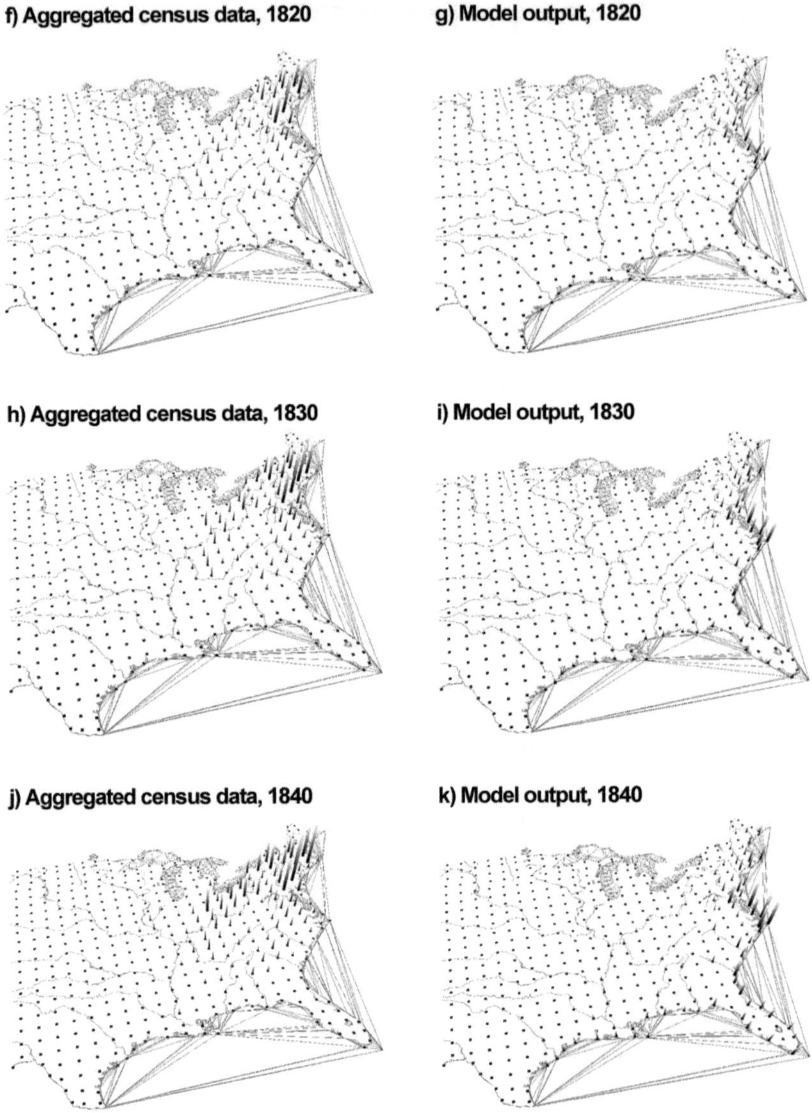

Fig. 10.8 (continued)

Each $W^{(k)}{}_i$ with $(k = 1, 2, 3, 4)$ could be taken as attractiveness factors for each of the four markets: grain, livestock/meat, lumber and manufactured goods; or it could be taken to represent different influences, such as level of economic development or technological advances – such as grain silos and 'booster' marketing of some of the cities.

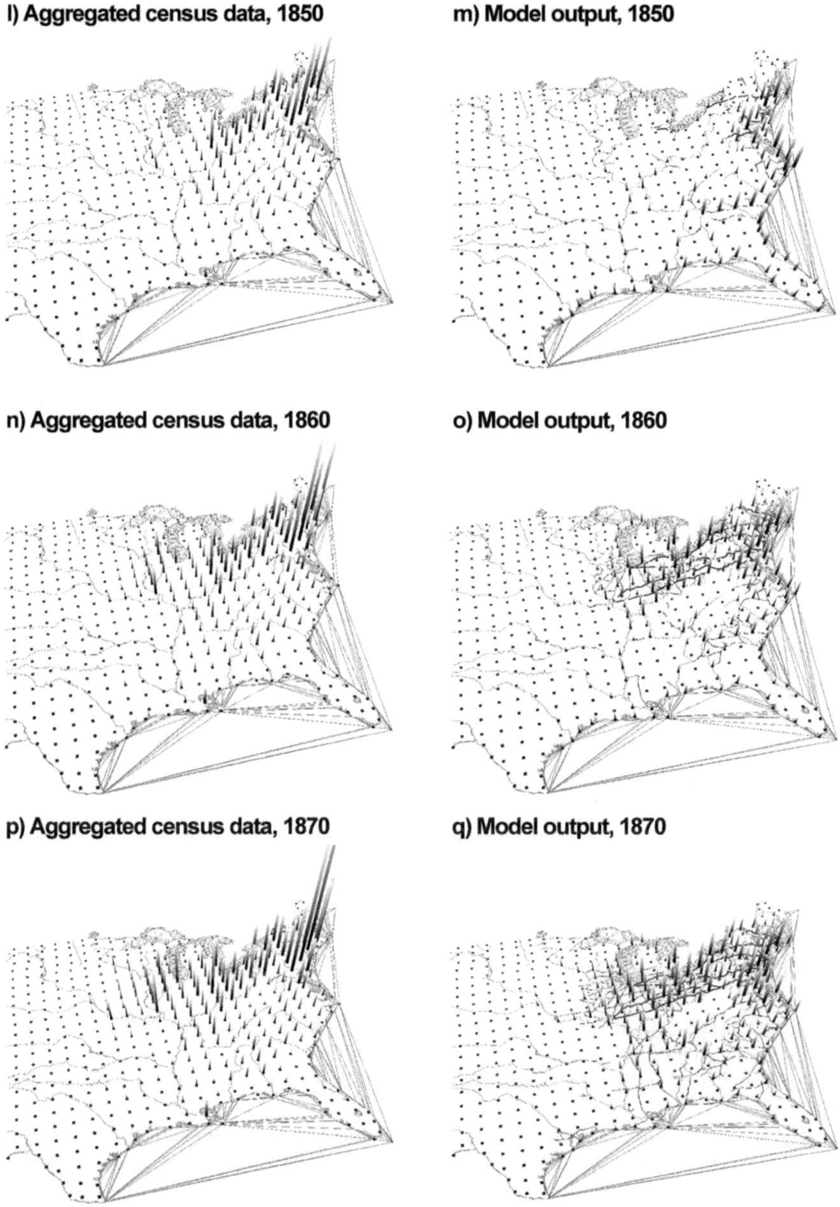

Fig. 10.8 Comparison of model with census data for each decade, 1790–1870

Transport System Dynamics

The railway lines have been entered exogenously for our model runs. More ambitiously, we could attempt to model the evolution of the transport network by adding a dynamic hypothesis which would be the transport system equivalent of Equation (10.7). A conjecture for this might be:

$$\Delta c_{ij} = \mu \left[S_{ij}(t+1) - S_{ij}(t) \right] \tag{10.20}$$

However, the task is more complicated than this. A real transport network is made up of links and a new railway line, for example, is a sequence of links in the network. Each link carries traffic from many origin-destination pairs and so there is an issue of whether we should try to reformulate Equation (10.20) on a link basis rather than an origin-destination basis.

Concluding Comment

The results that have been presented are offered on a 'proof of concept' basis: this is a potentially interesting way to explore historical geography. It is well known that the evolution of nonlinear dynamical systems is path dependent. We have introduced the idea of 'system DNA' as the sequence of initial conditions for the slow dynamics of our system of interest and we have paid particular attention to the changes in travel costs brought about by the introduction of railways. This shows how the acts of individuals – in this case the railway developers – influence the evolutionary path of the system of interest. Given that the model is a very simple and crude one, the fact that the results fit the data tolerably well suggests that this might be a rich seam for further exploration.

References

Berry, B. J. L. (1964). Cities as systems within systems of cities. *Papers of the Regional Science Association, 13*(1), 146–163.

Cronon, W. (1991). *Nature's metropolis: Chicago and the great west*. New York: W. W. Norton and Co.

Delaunay, B. (1934). Sur La Sphere Vide. *Izv. Akad. Nauk SSSR Otdel. Mat. Est. Nauk, 7*, 793–800.

Dijkstra, E. W. (1959). A note on two problems in connexion with graphs. *Numerische Mathematik, 1*, 269–271.

Harris, B., & Wilson, A. G. (1978). Equilibrium values and dynamics of attractiveness terms in production-constrained spatial-interaction models. *Environment and Planning A, 10*, 371–388.

Wilson, A. G. (1967). A statistical theory of spatial distribution models. *Transportation Research, 1*, 253–269.

Wilson, A. G. (1970). *Entropy in urban and regional modelling*. London: Pion.

Wilson, A. G. (2008). *Urban and regional dynamics – 3: Urban and regional 'DNA': The basis for constructing as typology of areas* (Working Paper 130). London: Centre for Advanced Spatial Analysis, University College.

Wilson, A. G., & Dearden, J. (2010). Phase transitions and path dependence in urban evolution. *Journal of Geographical Systems*. Springer Online First, DOI: 10.1007/s10109-010-0134-4.

Index

J. Stillwell, M. Clarke (eds.), *Population Dynamics and Projection Methods*, Understanding Population Trends and Processes 4, DOI 10.1007/978-90-481-8930-4,

MIX
Papier aus verantwortungsvollen Quellen
Paper from responsible sources
FSC® C105338

If you have any concerns about our products,
you can contact us on
ProductSafety@springernature.com

In case Publisher is established outside the EU,
the EU authorized representative is:
Springer Nature Customer Service Center GmbH
Europaplatz 3, 69115 Heidelberg, Germany

Printed by Libri Plureos GmbH
in Hamburg, Germany